Lecture Notes in Biomathematics

Managing Editor: S. Levin

W0106695

40

Renewable Resource Management

Proceedings of a Workshop on Control Theory
Applied to Renewable Resource Management and Ecology
Held in Christchurch, New Zealand
January 7 – 11, 1980

Edited by Thomas L. Vincent and Janislaw M. Skowronski

Springer-Verlag
Berlin Heidelberg New York 1981

Editors

Thomas L. Vincent
Aerospace and Mechanical Engineering
University of Arizona
Tucson, AZ 85705, USA

Janislaw M. Skowronski
Department of Mathematics, University of Queensland
St. Lucia, Queensland 4067, Australia

AMS Subject Classifications (1980): 49-XX, 69-XX, 92-XX, 93-XX

ISBN-13: 978-3-540-10566-4   e-ISBN-13: 978-3-642-46436-2
DOI: 10.1007/978-3-642-46436-2

Library of Congress Cataloging in Publication Data
Workshop on Control Theory Applied to Renewable Resource Management and Ecology
(1980: Christchurch, N.Z.) Renewable resource management. (Lecture notes in
biomathematics; 40) Bibliography: p. Includes index. 1. Conservation of natural resources--
Mathematical models--Congresses. 2. Renewable natural resources--Mathematical models--
Congresses. 3. Ecology--Mathematical models--Congresses. 4. Biological control systems--
Congresses. I. Vincent, Thomas L. II. Skowroński, Janislaw M. III. Title. IV. Series.
S912.W67 1980 333.7'2'0724 81-4515 ISBN 0-387-10566-2 (U.S.) AACR2

ACKNOWLEDGMENTS

This volume contains the proceedings of a workshop held January 7-11, 1980 at the University of Canterbury at Christchurch New Zealand. The workshop was co-sponsored by the U.S. National Science Foundation, the Australian Department of Science and the University of Canterbury.

We would like to extend our graditude to Professor R. S. Long and Doctor Hasha Sirisena for their role as New Zealand co-organizers.

PREFACE

As society becomes stressed by economic and population pressures, in turn, nature's renewable resources become stressed by harvesting pressures. For our own survival and euphoria, it is paramount that such resources remain as their name implies and not be driven to extinction through short term programs of over exploitation.

Consideration of the harvesting of renewable resources leads to a simple question that was the theme of the workshop and is the focus of these proceedings: Suppose you are assigned the role of manager for a specific renewable resource ecosystem. How would you decide on harvesting policies so that the system can be exploited economically yet at the same time maintain the integrity of the system? This, of course, is a loaded question. First of all, it is not clear that there is ever any one single decision maker who is able to set the rules for all of the harvesters in an exploited ecosystem. The political process is complicated and to some extent unpredictable. This aspect of the question is recognized to be important, but could not be addressed here.

Assuming then that someone really is in charge, what would be involved in the decision making process? As Clark[*] points out, "there is no alternative but first to model the system." We agree. However, if the original question was loaded, modeling is the adulterate. One can get a feel for this in the paper by Beddington, Botkin, and Levin. They give some ground rules for using models in resource management. They particularly address their comments to the applied mathematician or controls engineer, who is interested in management of biological systems. The appeal is to help develope models relevant and useful to the system in question. That is, models obtained from a detailed analysis of the particular ecosystem under study. Management decisions based on models "by hypothesis" (not necessarily justified) such as the Logistics and Lotka-Volterra models just aren't applicable in most management situations. The extent to which such models are not predictive for

---

[*] All references in the Preface are to papers contained in this volume.

a given ecosystem clearly measures the extent of inaccuracy for management purposes. Nevertheless a common practice has been to use simple single species logistic type of models with no age, size, or sex structure, with no account of parameter fluctuations such as carrying capacity for a multi-species situation where these factors may indeed be important. Such models are often used to calculate a maximum sustainable yield (MSY) for the system (which may be over estimated) and management decisions are based on a current comparison of population levels with respect to the MSY level. In spite of the reasonableness of this closed loop feedback approach, fisheries managed under this scheme have crashed often enough to invalidate the use of "simple" models in this fashion. Such models also largely leave out economic considerations. With productivity as the only economic factor the real value of the renewable resource is seldom realized.

Botsford addresses the task of formulating more realistic fishery models with the specific hope of drawing mathematical attention to more realistic models in the future. He includes age - or size - specific effects in the models. He also discusses how stability characteristics of the population can be measured in the fields. Levin's paper follows with models containing multiple age-spawning populations. Pulliam shows how to account for modification in animal behavior within a model. Goh addresses the problem of modelling and managing populations with widely variable recruitment.

On one hand more realistic modelling requires complexity of the models, and on the other hand there is rarely enough data available to build a complex model let alone a decent qualitative model. Complexity also requires sophisticated mathematics to handle age structure, spatial heterogenity, time delay, etc.. One alternative to complexity would be to tackle in a mathematical way system models with a lot of noise. One approach to this would be to form models flexible enough to search on-line for their own identity with the data available and at the same time assure stability of the system under harvesting and perhaps include other objectives. The latter problem is addressed by Skowronski with his Liaponov type adaptive identification conditions. Banks proposes an optimization-type technique (error minimizing) for the identification problem. He also discusses computational techniques associated with optimization of control systems.

One obvious aspect of management models which distinguishes them from other theoretical models used for prediction and/or understanding is that they must include a control input term to account for the effect of harvesting. While management models may indeed be drawn from population dynamic models, they are a distinctly different class of models by virtue of the added control input terms (i.e., management models can be driven about via the control inputs!). Population dynamic models can be analyzed for stability as discussed in the papers by Botsford, Levin, and Pulliam (where the fundamental question is how the introduction of more realistic terms effect the stability of the system) or they can be used for insight and predictive capabilities. By contrast, the analysis of management models cover this and much more. For example, because of the control inputs, the system can now be "optimized" with respect to some performance criteria. The performance criteria generally represents at least one economic input into the model. Clark's paper examines the consequences of including economic considerations in models for fishery management.

Another aspect of management models which differs from population dynamic models is in the concept of stability. For population dynamic models, stability is in reference to motion caused by a purturbation from an equilibrium condition. This concept must be expanded for management models as population changes are due not only to a combination of the inherent dynamics of the system, but are due to the control (harvesting) inputs as well. Only under a constant control input will there exist equilibrium points and will the stability concepts be applicable in the usual sense. Under non-constant control inputs the system is constantly being perturbed. Of interest here is a measure of the inherent stability of the uncontrolled system versus any destabilizing effects caused by the control inputs. A measure of this effect is given by the theory of controllable (or reachable) sets.

Its application to biological systems is in terms of the concept of vulnerability. The vulnerability of a biological system subject to harvesting is as central an issue as stability is for population dynamic models. Three papers in this volume are devoted to this subject. Vincent determines reachable set boundaries for a prey-predator system using optimal control theory with harvesting limits set by a game theoretic analysis of constant control equilibrium points. Fisher and Goh also use optimal control theory to determine reachable sets for several models of two species interactions. Grantham combines methods of controllability and Liaponov stability to devise a method for estimating reachable set boundaries that would be applicable for higher dimensional problems.

A number of papers are included that are case studies in which renewable resource management techniques are applied to specific ecosystems. Beddington examines changes in demographic parameters due to Krill harvesting in the Southern Ocean ecosystem and its effect on the population trajectories of whales. Putterill gives an appraisal of the extended New Zealand deep water fishery created by the new 200 mile exclusive economic zone. Comins and Trenbath examine strategies for optimal control of root knot nematodes in resistant-susceptible crop mixtures. McMurtrie gives a general model which describes how the availability of soil-water affects the photosynthetic growth of plants. The final paper in this volume by Rose deals with the one-dimensional transport of mobile nutrients in soils.

T. L. Vincent

J. M. Skowronski

Tucson, Arizona

St. Lucia, Queensland

October 1980

PARTICIPANTS

Dr. H. T. Banks. Professor, Division of Applied Mathematics, Brown University, Providence, Rhode Island 02912.

Dr. N. Barlow. Department of Botany and Zoology, Massey University, Palmerston North, New Zealand.

Dr. J. R. Beddington. Lecturer, Biology Department, University of York, Heslington, York YO1 5DD, England

Dr. D. B. Botkin. Professor, Environmental Studies Program, Department of Biology, University of California, Santa Barbara, California 93106

Dr. L. W. Botsford. Bodega Marine Laboratory, University of California, Bodega Bay, California 94923.

Dr. C. W. Clark. Professor, Mathematics, University of British Columbia, Vancouver, B.C., V6T 1W5.

Dr. H. Comins. Senior Research Fellow, Department of Environmental Biology, Research School of Biological Sciences, Australian National University, Canberra A.C.T. 2601.

Mr. R. F. Coombs. Fisheries Research Division, Ministry of Agriculture, P.O. Box 19062, Wellington, N.Z.

Dr. J. J. Doonan. Fisheries Research Division, Ministry of Agriculture, P.O. Box 19062, Wellington, N.Z.

Mr. G. Elvy. Department of Farm Management, Lincoln College, Canterbury, N.Z.

Mr. D. B. Esterman. Fisheries Research Division, Ministry of Agriculture, P.O. Box 19062, Wellington, N.Z.

Mr. M. E. Fisher. Mathematics, University of Western Australia, Nedlands, W.A. 6009.

Mr. R. I. C. Francis, Fisheries Research Division, Ministry of Agriculture, P.O. Box 19062, Wellington, N.Z.

Dr. B. S. Goh. Senior Lecturer, Mathematics, University of Western Australia, Nedlands, W.A. 6009.

Dr. W. J. Grantham. Associate Professor, Mechanical Engineering, Washington State University, Pullman, Washington 99163.

Mr. T. Halliburton. Electrical Engineering, University of Canterbury, Christchurch, N.Z.

Dr. L. S. Jennings. Mathematics, University of Western Australia, Nedlands, W.A. 6009.

Mr. D. Johnstone. Mathematics, Massey University, Palmerston North, N.Z.

Dr. M. A. Jorgensen. Biometrics Section, Ministry of Agriculture, P.O. Box 1500, Wellington, N.Z.

Dr. S. A. Levin. Professor, Biological Sciences, Ecology and Systematics, Langmuir Laboratory, Ithaca, NY 14850.

Dr. M. K. Mara. Applied Mathematics Division, D.S.I.R., P.O. Box 1335, Wellington, N.Z.

Dr. C. McLay. Zoology, University of Canterbury, Christchurch, N.Z.

Dr. R. McMurtrie. CSIRO, Division of Forest Research, P.O. Box 4008, Canberra, A.C.T. 2600.

Dr. R. H. Pulliam. H. S. Colton Research Center, Museum of Northern Arizona, Flagstaff, Arizona 86001.

Dr. M. Putterill. Department of Accountancy, University of Auckland, Private Bag, Auckland, N.A.

Dr. C. Rose. Professor, School of Australian Environmental Studies, Griffith University, Nathan, Brisbane, Queensland 4111.

Dr. J. C. Rutherford. Ministry of Works and Development, Private Bag, Hamilton, N.Z.

Dr. H. R. Sirisena. Reader, Electrical Engineering, University of Canterbury, Christchurch, N.Z.

Dr. J. M. Skowronski. Reader, Mathematics, University of Queensland, St. Lucia, Brisbane, Qld. 4067.

Dr. A. C. Soudack. Professor, Electrical Engineering, University of British Columbia, Vancouver, B.C. V6T 1W5.

Dr. R. J. Stonier. Mathematics, University of Queensland, St. Lucia, Brisbane, Qld. 4067.

Dr. A. C. Tsoi. Senior Lecturer, Electrical Engineering, University of Auckland, Private Bag, Auckland, N.Z.

Dr. T. L. Vincent. Professor, Aerospace and Mechanical Engineering, University of Arizona, Tucson, Arizona 85721.

Dr. D. J. Wilson. Lecturer, Mathematics, University of Melbourne, Parkville, Victoria 3052.

CONTENTS

CONTENTS (continued)

MATHEMATICAL MODELS AND RESOURCE MANAGEMENT

John Beddington
Biology Department
University of York
Heslington, York, England

Daniel Botkin
Biology Department
University of California
Santa Barbara

Simon A. Levin
Biological Sciences
Cornell University
Ithaca, New York

The fact that resource management regimes are breaking down in almost all ecological applications dramatizes the need for a technology based on an adequate theory. Failures in pest control and fisheries bear grim witness to this need.

While many new pesticides, which are often target specific, have been developed, the increase in pesticide-resistant arthropod species over the last 15 years has been dramatic: pesticide-resistant species have more than doubled in the period 1965-1977 (Figure 1).

Many of the major fisheries of the world have suffered severe depletion in stocks, and yields have declined. In some cases, an extreme population 'crash' has ensued from inadequate or incompetent management, management based on erroneous or incomplete theory or a combination of these and as yet unknown factors. The success story of the south-east Pacific Anchoveta during the 1960s suddenly turned sour as catches fell from the peak of 13.1 million tonnes in 1970 to only 1.2 million tonnes in 1978. The North Sea herring catch of 1978 was only 2% of the 1965 peak catch (30,000 tonnes and 1,469,000 tonnes respectively). The north-west Atlantic catches of cod decreased from 1,877,000 tonnes in 1968 to 483,000 tonnes in 1978; a decline of nearly 75%. The north-west Atlantic mackerel catch declined to 7% (1978) of peak catch in only 5 years and capelin yield in the same area in 1978 was only 25% of the peak of 367,000 tonnes in 1975. (FAO, 1979; Miller and Carscadden, 1979).

Similar declines in such major fisheries as the South African pilchard, the haddock, the north-west Atlantic herring, the Norway pout and many others may be viewed through the FAO yearbooks of fishery statistics.

In considering these problems, it is essential to distinguish two roles for models: the first a purely metaphorical one, meant for suggestion and understanding,

the other a management-oriented one, with prediction the objective. In the former
category, one should place most of the vast literature in theoretical ecology.
Many theoretical papers derive conclusions which might in some cases be used as
general guides to management, but the objective rarely is to produce a specific
literal model, and it is a serious error to fail to recognize the distinction. For
example, the logistic equation has been used to make clear the principle that a
competitive bioeconomic system will not reach equilibrium at a maximum sustained
yield for an exploited renewable resource as in the case of fisheries or whaling;
however, to go beyond this heuristic use and to base management policy upon the
logistic or other equally over-simplified models is to misunderstand the above and
can lead to management catastrophes. One such example is the case of the north-
west Atlantic capelin whose catch limits were set on the basis of the 'surplus'
capelin available following depletion of its major predators (Winters 1975; Winters
and Carscadden, 1978). The dynamic consequences of harvesting at a level determined by such a procedure may be simply investigated and, in fact, have a high chance
of leading to prey depletion (Beddington and Cooke, 1979). It is, indeed, worrying
that much discussion of krill potential yield has centred on similarly naive calculations. Such an over-simplified model necessarily - indeed by design - ignores a
great deal: interactions with other populations; age, size and spatial structure of
the population; and the inherent uncertainty in the system. As the real situation
becomes more complex, the inapplicability of simple models becomes more apparent.
Models of interacting populations of aggregated type in which population size is
represented by a scalar, for example the Lotka-Volterra equations, can in almost
no situations be validated and results based on them are purely illustrative. As
instruments for suggestion or communication of ideas, they can be very effective
and powerful; however, to construct a management strategy which placed explicit dependence upon these models would be foolhardy.

At this point, the control engineer might well ask, "Okay, if the Lotka-
Volterra models are not the right ones to use, then what are the right ones? Just
tell me, and I will bring my methods to bear upon them." This is not the appropriate question, and indicates why it is essential to involve the modeller with the
analysis of the biological system in a fundamental way, as early as possible in the
study. Unlike physics, population ecology does not have a set of 'Newtonian laws'
on the basis of which population dynamics may be understood. The same can be said
about ecosystems and ecosystem dynamics. Ecological axioms are hypotheses and nothing more; unfortunately, no matter how carefully restrictions and assumptions are
spelled out in the initial presentatation of the model, it is virtually impossible
to prevent usage of the model beyond its tolerance limits, and this is why the control scientist must dig below the surface and not accept the model as given. Moreover, the modelling process is itself an activity in which the skills of the mathematician or engineer can be most useful, and these individuals must overcome their

reluctance to become involved at this earlier, sloppier stage.

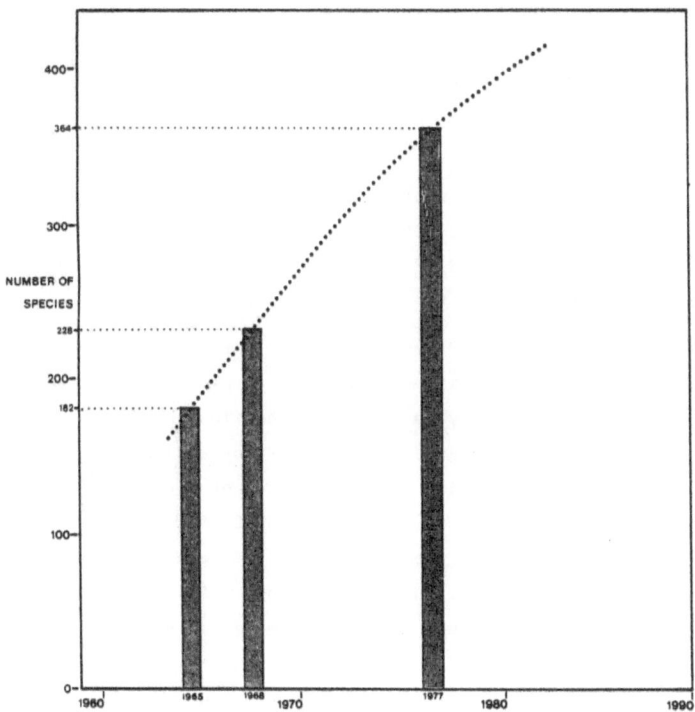

FIGURE 1.   Number of pesticide-resistant species of insects and mites over time.
(from FAO, 1977. Report of the first session of the FAO Panel of Experts
on Pest Resistance to Pesticides and Crop Loss Assessment.
FAO Plant Production and Protection Papers, 6.)

Another reason that it is fruitless to hope to be handed the model on a silver
platter is that there is no single such model. First, one must settle the level of
detail necessary for description, and this cannot be answered in the abstract; it
must depend upon the purposes to which one wishes to apply the model. Most ecological phenomena exhibit stochasticity, time delays, and distributed characteristics
related to age structure, sex and reproductive classes, spatial position, and behaviour. Parameters are not fixed but are subject to change. In general, inclusion of
all these features would result in a model so cumbersome that it could not be reli-
ably used, but ignoring key ones renders the models impotent. The decisions as to
the appropriate level of structure and detail involve both biological and mathematical expertise, generally requiring a well-integrated team of scientists.

Once the level of detail is determined and the questions to be asked clarified, the problem of the correct description remains. Proceeding only with phenomenological models, treating the systems as black boxes, is at best a risky business and can be justified only when other information is not available. In developing models at one level of complexity, it is generally best if one understands how things are functioning at lower levels: thus population models must be constructed on the basis of known physiological, anatomical, and behavioural functional dependencies, and both field and laboratory work are indicated. Much work in modelling has lacked this crucial component.

In summary, it is unclear what level of generalization is possible in dealing with ecological systems, and it will be frustrating to try to develop a modelling strategy applicable to, say, all fisheries. Case by case investigations are essential, as is the willingness to settle for restricted models.

There is an important, indispensable, role for modelling in the development of management strategies, but it is essential not to place too much emphasis on particular models. When in the analysis of a model new insights are exhausted, then it is time to quit; the tendency to squeeze models dry by analyzing every detail out of them, computing every equilibrium and every basin of attraction, is of neither ecological nor mathematical interest. Modellers must emphasize the robustness of their conclusions, and concern themselves with the sensitivity of their models to changes in structure or detail; those conclusions which are seriously model-dependent are of little general value. Both in viewing one's own models and those of others, it is profitable to ask the troublesome question, "What is the least serious way the model can be changed and still destroy the results?"

Not only the models themselves, but also the paradigms of ecological theory require scrutiny. For example, past attempts often have been based, explicitly or implicitly, on a notion of stability equivalent to the stability of a mechanical system with a single stable equilibrium point, despite the existence of ample evidence of change and lack of stability in this restricted sense in ecological systems. The concept of ecological stability still requires revision, and it is certainly not axiomatic that the relevant concepts of stability for management will be identical to those most appropriate for scientific understanding: the distinction involves political, economic, and aesthetic determinations.

Above all, the modeller must avoid coming with a bag of tricks and essentially asking the biologist where they apply. The engineer or mathematician, as the biologist, must be problem-oriented and allow the methods to define themselves. Again, this involves profound and early involvement, but only in this way can the modeller judge when a model is a good one, and what kinds of simplifications are possible. Success stories in modelling all involve collaboration and involvement. Ecosystems are sloppy systems, with much noise, and with immigrations, emigrations, extinctions, and environmental change. It is not clear that standard engineering mathematics can be applied without major alterations to give either theoretical insight or practical benefits, but the development of the relevant modifications involves stimulating challenges for modellers.

The manager is faced with a dilemma: he cannot project optimal or reasonable solutions without a model, and if appropriate models don't exist, he is at a loss as to how to proceed. The first step is the recognition that biological systems are not machines or factories, and that they are organized according to different principles than mechanical devices. We need to develop techniques and methodologies, in some cases borrowed from existing areas of mathematics such as stochastic optimal or adaptive control or statistical decision theory, which can deal with systems in which random effects are important. In some cases, entirely new methodologies will be required. We need strategies to deal with the management of poorly-specified systems in which not only the parameters of the models, but their very forms are in question. An appropriate theory for management does not yet exist in part because control theories have been based on models not very selectively borrowed from theoretical ecology, rather than on ones developed explicitly for management. It is to be hoped that the current drift in interest among modellers from more theoretical questions to management will lead to the development of a more robust science.

REFERENCES

Beddington, J.R. and Cooke, J. 1979. Harvesting from a Predator-prey System. Manuscript submitted to Math. Biosc.

FAO. 1979. Yearbook of Fishery Statistics, Vol. 46. Catches and landings, 1978

Miller, D.S. and Carscadden, J.E. 1979. Biological characteristics and biomass estimates of capelin in ICNAF Division 2J and 3K using a sequential capelin abundance model. ICNAF RES. DOC. 79/11/32 (revised)

Winters, G.H. 1975. Review of capelin ecology and estimation of surplus yield from predator dynamics. ICNAF RES. DOC. 75/2. Ser No. 3430

Winters G.H. and Carscadden, J.E. 1978. Review of capelin ecology and estimation of surplus yield from predator dynamics. ICNAF RES. BULL. 13: 21-30

MORE REALISTIC FISHERY MODELS:
CYCLES COLLAPSE AND OPTIMAL POLICY

Louis W. Botsford*
Bodega Marine Laboratory
P.O. Box 247
Bodega Bay, California 94923

Simple classical approaches to mathematical analysis of fisheries are
inherently limited. Increased use of more realistic population models,
in the form of age- or size-specific, density-dependent models based on
physiological energetics, is proposed here. Recent, fishery-related
results using models of this type illuminate such issues as cyclic behavior of populations, multiple equilibrium levels, and optimal policy.
Specific results include: size-selective fishing can lead to unstable
cycles, an increase in individual growth rate can maintain depressed
equilibrium levels, optimal policy for size-specific, density-dependent
models may involve pulse-fishing, and more realistic models of multi-
species problems in the Antarctic may alter conclusions reached through
simpler models.

INTRODUCTION

The increasing importance of food resources and the questionable past record
of fishery management imply better management techniques are required for the future. While much effort has been expended toward increasingly sophisticated analysis of existing mathematical models of fisheries, very little has been applied
to improving the models themselves. Age- or size-specific models based on concepts
of physiological energetics may provide better fishery models. The rationale behind use of these models in fisheries, some of the models of this sort currently in
use, recent results obtained through their use and how they might apply to multi-
species management of Antarctic fisheries are discussed in this paper.

Existing fishery models (cf. Ricker 1977, Gulland 1977) are of three basic
types that have existed in the same form for twenty years or more: the logistic
model (Graham 1935, Shaeffer 1954), the dynamic pool model (Beverton and Holt 1957),
and the stock-recruitment model (Ricker 1954). Each of these falls short of a complete, realistic description of population dynamics. The logistic model, for example, reflects the self-limiting behavior of population growth, but does not include the mechanics of growth, reproduction and mortality that account for this
behavior. The dynamic pool model focuses on growth and mortality of individuals
(assumed to be identical and not density-dependent) while reproduction is not

---

* Current Address: Department of Wildlife and Fisheries Biology, University of
California Davis, CA   95616

accounted for. The stock-recruitment model, on the other hand, focuses on density-dependent reproductive processes and survival of the young at the expense of a description of growth and mortality at older ages. In addition to these three basic types of models, the latter two have been combined to provide a more complete view of fish populations (cf. Beverton and Holt 1957, Walters 1969). The behavior of these combination models and optimal fishery policy using them are usually examined through simulation with only rare attempts at further analysis (cf. Getz 1979).

Formulation of better models requires definition of what makes a model better or worse. Models can be viewed as inductive arguments by analogy. The basis of this argument is simple inductive enumeration. Since the model behaves the same as the object modeled in ways A, B, C, it will behave the same in way D. For example, A, B, C could correspond to time periods in the past and D to a time period in the future. Arguments by analogy become stronger as the number of instances of enumeration of relevant similarities increases (Salmon 1973, p. 97), or in other words as the realism of the model increases.

With regard to population models this implies that models become better as the number of ways in which they match reality increases. However, the number of similarities between models and populations could be increased in several ways. The most obvious is to match population level behavior over as long a time period in the past as possible in order to predict the future. A second way that is possibly implicit in use of the logistic model is to choose a model that matches behavior of other populations and conclude, therefore, that it will match the behavior of the population of interest. For example, the logistic model fits laboratory growth of Escherichia coli, Paramecium spp., Drosophila sp. and Moina sp. populations (Hutchinson 1978), therefore it will reflect the dynamics of exploited tuna populations. These two possibilities appear to exhaust the possible arguments by enumeration in support of models that are based on behavior at the population level.

However, there are additional instances of enumeration that in many cases can be based on relevant existing knowledge. Behavior at the population level, the level of interest in fisheries, is determined by behavior at the level of individuals within the population; that is individual growth, mortality and reproductive rates. In many cases these rates are known or can be estimated from existing data. Thus, models can match the real population in terms of individual behavior rather than only behavior at the population level. Inclusion of information regarding individuals provides the potential for better fishery models in the sense that it provides stronger inductive arguments by analogy.

As models become more realistic, they could become hopelessly detailed, requiring prodigious quantities of data. How far should this process be carried? Part of the answer to this question is that the model be at least of sufficient detail to include the mechanisms responsible for all behavior of interest. A second part of the answer is that models should be complex enough to allow use of existing data of relevance (e.g., data concerning individual growth, mortality and reproductive rates). In many practical applications simple models are used because of the alleged paucity of data. An alternative approach that appears more reasonable would be to examine the effects of having more data and then to decide whether to gather them on an economic basis (see Discussion). The point here is that current fishery models are oversimplified so that in many cases they do not include all available data, cannot exhibit the behavior observed in real populations, and tend to lead a complacent attitude regarding how much is actually known about a population.

In the remainder of this paper I discuss how age-, size-specific models may better serve fishery analysts and managers. I will show that these models represent population behavior more realistically than classical fishery models. In doing so, I hope to draw more mathematical attention to these models in the future.

AGE-, SIZE-SPECIFIC MODELS

Age-, size-specific models are those that "keep track" of the number of individuals of each age or size (or both) in a population. These models appear in several different forms in the ecological literature depending on the discrete or continuous nature of time and age, and on which variables are of primary interest. The Leslie matrix was an early discrete time, discrete age population model (Leslie 1945). The widely-used concepts of fertility tables and life tables is a continuous time version of this model (Lotka 1925). A continuous-time, continuous-age version of this model, the Von Foerster equation (or in mathematical physics, the continuity equation), was introduced as a population model by Von Foerster (1959). Sinko and Streifer (1967) developed a similar model in terms of continuous time and both age and size. Changes in the number of animals at each size and age with time were determined by growth, mortality, and reproductive rates which in turn depend on age, size, density, and environment.

An immediate problem in the use of models in fisheries is deciding which variables associated with individuals (e.g., age, size) are necessary to describe population behavior. This decision can be based on two considerations: (a) choose the individual variables that are of commercial significance, and (b) choose those individual variables that most critically affect population dynamics. With regard to the first, the individual characteristic of most importance commercially is individual size (length or weight). With regard to the second, dynamic behavior of populations is determined by individual growth, reproductive and mortality rates. A review of these vital rates in exploited aquatic populations has shown that they usually depend heavily on size, less closely on age and also on an inherent variation among individuals (Botsford 1978). Ultimately, the choice of variables depends on the nature of vital rates for the specific population being modeled.

Most of the results for this class of models have been obtained for age-specific models. However, in most aquatic individuals vital rates depend on size rather than age. In addition, growth rate, at least in juveniles, is density dependent. These two conditions require a model that at least includes a size variable to describe population behavior.

BEHAVIOR AND OPTIMAL POLICY

While age-specific models have been in use for many years, very little is known about behavior and optimal fishery policy for realistic non-linear age-, size-specific models. This section is a qualitative description of the salient features of some recent results for age-, size-specific models. Local stability of an age-specific model with density-dependent recruitment is discussed first to provide the basis for a global analysis of an age- and size-specific model with density-dependent recruitment and growth rate. These are followed by some results regarding optimal fishery policy for a general density-dependent, size-specific model. Ecologically relevant behavior of these models is stressed rather than a complete view of the mathematical characteristics.

Behavior of linear versions of age-specific models (e.g., the Leslie matrix) is well known (Lotka 1925, Leslie 1945, Keyfitz 1978). The dominant solution is an exponential reproductive rate that increases with time (unstable about zero) or decreases with time (stable about zero) (or under fortuitous circumstances remains constant). Since the essential mechanisms that prevent populations from going to zero or infinity are lacking in this model, its behavior is unrealistic. It is, therefore, of limited usefulness in practical problems.

Behavior of non-linear age- or size-specific models has been analyzed for only a few cases of interest. For age-specific models with density-dependent recruitment, early results were obtained through simulation. Ricker (1954) established the fact that the slope of the stock-recruitment curve at the replacement point had to be greater than minus one for stability of a single age-class population. He then noted that in simulations of multiple age-class populations stability was possible when this slope was less than minus one. A second observation by Ricker (1954) that was later supported in other simulations by Allen and Basasibwaki (1974) and Menshutkin (1964) was that fishing made populations more stable.

Botsford and Wickham (1978), in an attempt to better understand cycles in the northern California Dungeness crab fishery, investigated these two issues further. Their model is a continuous-time, continuous-age, age-specific model in which recruitment was the product of reproduction and a survival function that reflects survival to recruitment. Each of the latter two depends on a weighted sum (integral) of older individuals in the population. This model is similar to those of Ricker (1954) and Allen and Basasibwaki (1974) except that reproductive rate and density-dependent survival do not necessarily depend on the same weighted sum over older age classes. There is no a priori reason to expect that the mechanism responsible for density-dependent recruitment (e.g., cannibalism) will depend on older individuals in the same way that fecundity does.

Equilibrium conditions were determined and local stability was analyzed for the model linearized about equilibrium. Stability conditions are in a form that illuminates the forementioned multiple age-class and fishing versus stability issues. Stability for the compensatory mechanisms of interest depends on two numbers: K, the "normalized" slope of the recruitment survival function at equilibrium, and K', a number that depends on the relative influence at each age of older animals on recruitment (e.g., fecundity and propensity to cannibalize at each age). The model is stable for K > K' (decreasing oscillation) and unstable for K < K' (increasing oscillation).

Since K depends on f and the equilibrium level, it doesn't change with the addition or removal of older age classes (unless equilibrium level changes). The value of K', however, responds more directly to changes in age structure. Values of K' for a specified model can be obtained numerically (K' is the value of K for which the real part of the dominant eigenvalue is zero). From numerical solutions under several different conditions and analytical solutions for two cases, the behavior of K' can be roughly described. The value of K' corresponds to Ricker's (1954) result for the single-age-class case (a slope of -1). As the number of age-classes increases, K' tends to decrease. An approximate solution for K' showed that it increased with the ratio of the mean age of influence of older animals on recruitment to the standard deviation. Thus a narrow, peaked influence (of older animals on recruitment) over age is generally less stable than a broad, flat influence.

These results support the observation that the slope of the stock-recruitment curve can be less than minus one as the model changes from a single to a multiple age-class model. However, they do not support the earlier conjecture that fishing stabilizes populations. Age- or size-selective fishing removes animals of older ages, which narrows the influence over age of older animals on recruitment and can make the populations unstable. For example, in several cases of a model of the northern California Dungeness crab population size-selective fishing caused K' to become more positive, thus making the population inherently less stable.

In addition to the fact that fishing can make populations unstable, a second result of this work is of practical importance. Since stability criteria can be simply expressed in terms of K and K', stability characteristics can be measured in

the field. We are currently engaged in comparison of different biological mechanisms (e.g., cannibalism, an egg-predator worm whose population size is proportional
to the number of females, intra-specific competition for resources, etc.) to determine which is responsible for the cycles in the northern California crab fishery. To determine whether or not each of these is the cause of cyclic behavior,
we are essentially determining values of K and K'. Of course, the actual mechanism
of cycles is more complex than a simple linearized analysis would indicate, but
nonetheless the change in survival with density and relative influence of older
individuals on recruitment are key parameters to be determined.

Related analyses of similar models in a fishery context include those of
Allen and Basasibwaki (1974) and Levin and Goodyear (1979). Allen and Basasibwaki,
using a discrete time and age version of the above model derived stability conditions for some cases. Levin and Goodyear (1979), also using a discrete time and
age model with a Ricker stock-recruitment relationship, have examined the effect
on stability of mortality and reproduction at older ages. An important result is
that as mortality rate of older individuals is increased, it can first increase,
then decrease stability. Thus even fishing that is not age- or size-selective can
make a population less stable. (This result differs from a statement in Botsford
and Wickham (1978) to the effect that non-selective fishing always stabilizes a
population. This statement is in error.)

There have been far fewer attempts at analysis of global stability of density-
dependent, age- or size-specific models. Botsford (1980a), in an attempt to explain
the protracted decline of several exploited populations, analyzed a model similar
to the above model of Botsford and Wickham (1978) except that it was size-specific
and included density-dependent growth rate. The populations of specific interest
were the Dungeness crab population of central California, the Eurasian perch population of Lake Windermere in England and the Pacific sardine off California. These
populations have three characteristics in common: (a) following exploitation, a
decline to low levels from which they have not recovered, (b) an increase in individual growth rate during or following the decline, and (c) a compensatory effect
on recruitment that depends on older members of the population.

From the results discussed previously both equilibrium and stability depend
on the relative influence (on recruitment) of older animals at each age. The actual biological influence on recruitment probably depends on size rather than age.
For example, fecundity and cannibalism would most likely depend more on size than
on age. If the actual mechanism were size-dependent and the growth rate changed,
then equilibrium and stability could change appreciably.

Botsford (1980a) analyzed several aspects of global stability. He determined
that an increase in growth rate could lead to lower equilibrium recruitment. This
lower recruitment level was due to individuals reaching "cannibalistic size" in two

years at higher growth rate rather than four years at lower growth rate. If they are exposed to constant mortality pressure during each of these years, more of them survive to "cannibalistic size" (per recruit) at the higher growth rate.

After establishing the possibility of different equilibrium recruitment rates corresponding to different individual growth rates, a model of the central California Dungeness crab population was simulated to see if populations could actually switch from one equilibrium level to another. The simulations included gradually increasing fishing mortality for two kinds of exploitation: in one only males are fished and in the other both males and females are fished. For the males-only case, the fished system could be pushed into a different domain of attraction, corresponding to low equilibrium recruitment and high individual growth rate, by several years of poor recruitment. That the unfished population was less susceptible to this input supports the contention of Murphy (1968) regarding relative imperviousness of iteroparous and semelparous populations to environmental fluctuations. For the case in which both males and females are fished, this same type of behavior was observed. However, an additional type was also present: the gradual increase in fishing pressure eventually led to a point at which recruitment decreased and individual growth rate increased in a regenerative fashion until a new equilibrium was reached at low recruitment and high growth rate.

Whether or not this kind of mechanism is actually the cause of depressed levels remains to be determined. However, the main point here is that it is a possible mechanism that has not yet been considered as a potential cause of these declines. Furthermore, illumination of this mechanism is not possible without age-, size-specific models. An important aspect of this mechanism is that it reflects earlier comments on resilience of ecosystems (Holling 1973). Here is a possible example in which man has reduced populations to a low level from which recovery will be diffiult, it not impossible.

Optimal harvest policy of age- or size-specific models has been addressed largely with the use of the linear Leslie matrix model. The results obtained are of limited value because of the linearity of the model. Because the model is linear, only a population with dominant eigenvalue greater than one (i.e., an increasing population) could yield a sustained harvest. Since the population is increasing exponentially, optimal policy involves waiting until just before the end of the planning period, then harvesting the whole population. The optimal harvest problem with this model has been made more reasonable by appending constraints (such as maintaining a constant level) (Beddington and Taylor 1973, Rorres and Fair 1975, Rorres 1976). The limitations of this approach and this model are discussed in Beddington and Taylor (1973), Mendelssohn (1976), Reed (1979), and Botsford (1980b). More realistic results can be obtained by including the actual biological "constraints" on population growth rather than attaching artificial constraints to an unrealistic model.

Botsford (1980b) formulated the optimal harvest problem in terms of a size-specific model with a density-dependent growth and recruitment rates. These rates could depend either on a weighted sum of other individuals in the population (similar to the crab model discussed above) or on a food variable that in turn depended on consumption by the population. Necessary conditions for values of size limits and fishing effort that maximized discounted future profits were determined. These results reduced to those obtained by others for the special cases of the linear, age-specific model and the single age class model.

The resulting necessary conditions were similar to those obtained by others, but extended earlier results in the direction of realism and completeness. Clark et al. (1973) described optimal policy for a Beverton-Holt model of single age class as a balance between the rate of increase in value of stock due to "economic" reasons (i.e., discount rate times net biovalue) and the rate of increase in value of a population due to biological causes (growth and mortality rates for the age class). A corresponding expression can be interpreted similarly for optimal policy using the logistic model, but the biological variables involved (r and K) are not as easily associated with real populations as individual growth and mortality rates. The former model is not complete in that it lacks a self-sustaining, reproductive process, while the latter is vague and less realistic. The model used in Botsford (1980b) extends the realism of the former to include the density-dependent, self-sustaining nature of the latter. The corresponding necessary condition is again a balance between the discount rate times net biovalue and rate of increase of stock due to biological factors. However, in this case the latter is the sum over sizes to be fished of rates of increase in biovalue due to individual growth, reproductive and mortality rates, as well as the negative values due to food consumption and density-dependent effects of older animals. Decisions made on the basis of this expression include a much more detailed consideration of individual growth and metabolism. For example, the effect of changes in metabolic efficiency with age (Paloheimo and Dickie 1965) and the effect of changes in relative market value versus future reproduction on harvest policy can be evaluated.

The results also showed that optimal policy solutions differ radically from those obtained with simpler models. For example, optimal policy for the logistic is (for certain parameter values) a policy of constant effort and population level (Clark 1976). However, when the realism of explicit multiple age classes with density-dependent growth and recruitment rates are included in the problem, optimal harvest policy is no longer a constant policy. Optimal policy probably involves pulse fishing. This kind of policy is not without precedence. Clark (1976) in analyzing a constant-recruitment, multiple age-class model made the conjecture that pulse fishing was optimal. Several others have obtained this kind of result in simulation. Hannesson (1975) examined the case in which fishing costs are not zero, and concluded that pulse fishing was slightly better than the stationary fishing policy tried. Walters (1969) numerically obtained the maximum yield of models with individual growth and stock-recruitment relationships for two cases: one with and one without fishing selectivity. In the former case, a constant policy was optimal while in the latter a pulse fishing policy was optimal. Pope (1973), using a similar model with no gear selectivity, also numerically determined optimal yield to be a policy involving pulse fishing. Getz (1979), also using a model that includes a stock-recruitment relationship and individual growth, recently obtained a numerical solution for maximum yield with a description of harvesting similar to that used here; constant fishing mortality for all sizes greater than a minimum size to be determined. In his solution fishing mortality and minimum size were assumed constant with time.

ANTARCTIC KRILL, BALEEN WHALE FISHERIES

Multispecies fisheries present problems that have not yet been addressed in terms of age- or size-specific models. May et al. (1979), in their analysis of the Antarctic fisheries based primarily on population-level, logistic-type models pointed out the necessity of bringing more complex models to bear on these problems. In this section, several probable differences between the results of May et al. (1979) and results of an age- or size-specific modeling approach based on the behavior of individuals are discussed.

Consideration of age-specific characteristics of individuals would probably change the estimate of how much krill is available for harvest. The current estimate, 150 million tons surplus krill per year, is based on estimates of food consumption rates based on physiological energetics of whales (Lockyer 1972) and estimates of the difference between the total whale biomass before exploitation and present biomass (Mackintosh 1973, Laws 1977). Basing the estimate of surplus krill on a biomass deficit and percentage consumption rate per unit biomass may incur significant error. Since the age and size structure of the population have changed with exploitation, the total annual metabolic requirement per unit biomass may also have changed. This is due to two causes. One result of exploitation is a shift in size distribution toward smaller sizes. For most aquatic animals the rate of food consumption per unit body weight is greater for smaller, younger individuals and decreases as the individual grows older. Thus a change in size structure would change consumption rate per unit weight. A second cause of change in consumption rate per unit biomass is increased consumption rate itself. When fish are provided a greater ration of food, they often eat more (up to a point) and grow faster. Whales also grow faster in response to greater food availability. In addition, they show earlier maturation and exhibit higher pregnancy rates. Both Fin and Sei whales have been observed to mature at approximately half the age at which they matured prior to exploitation (Lockyer 1972, Lockyer 1974) and higher pregnancy rates have been observed in these two species and the blue whale (Laws 1977). It is likely that these whales are consuming more food per unit weight than they were earlier.

A second way in which age- or size-specific models would differ from the simpler approach is in the actual behavior of the models themselves. These could be significant, qualitative differences. For example, in equation (1) of May et al. (1979) the rate of decrease in krill due to predation is proportional to the product of sizes of the two populations. In actual fact, however, the consumption rate of an individual whale is not based on the total number of krill in the Antarctic, but rather on the density of krill available to him (or her). The functional dependence of consumption rate on prey density would probably increase with prey density to a maximum rate (as in Ivlev 1961). The point here is that at krill levels that are not extremely low, the last term in equation (1) of May et al. (1979) would be proportional only to whale population size. The effect of this change on results could be a considerable change in static and dynamic aspects.

A third probable difference in results is in dynamic behavior of the system. At least some whale populations apparently respond to changes in food level by changing reproductive parameters. The response of the population (a change in population size) to a change in environment (food level) occurs after a lag equal to the time required for an increase in reproduction to be felt as an increase in population of whales of significant size (i.e., adults). With increasing exploitation of whales, this lag is decreasing, hence response time or "characteristic return time" is decreasing (see May et al. 1978, p. 241 for a discussion of this phenomenon). This effect would act approximately in opposition to the mechanisms that lengthen response time discussed in May et al. (1979).

A fourth difference that is more qualitative than the others is simply an increased emphasis on measurement of results at the level of the individual rather than at the population level and inclusion of these in management decisions. Recent changes in individual whales and individuals in predator populations (e.g., the lower age of sexual maturity of crab-eater seals (Laws 1977) can be evaluated quantitatively in terms of their effects on prey populations (consumption rates) as well as their own population dynamics. The International Whaling Commission currently includes age structure in their population models (Allen and Kirkwood 1977). Perhaps these could be expanded to include multi-species considerations. These could be based, at least partially, on current knowledge of physiological energetics of whales (Lockyer 1972, Brodie 1976).

In summary of this section, it is impossible to definitely state at this time that simple models will lead us astray in the Antarctic. However, there are additional data available that could be used, and an approach that involved more realistic models would emphasize gathering more data at the individual level. The

important point here is that, while logistic-type models may provide convenient
metaphors, we should not allow them to lull us into such a complacent frame of mind
that we believe that they realistically predict the future response of this system.
For example, we have removed a predator and competitor from the Antarctic (whales),
and prey abundance as well as competitor abundance (e.g., seals) have increased.
This state of affairs may tempt us to predict that if we were to stop fishing whales
the system would return to its previous state. Whether or not it does is going to
depend on predator-prey and competitor interactions at the individual level. Accompanying any new changes in our effects on the system with detailed monitoring
of changes in age-, size-structure and the characteristics of individuals seems a
better policy than changing policy, then waiting to observe population level effects.

DISCUSSION

The above examples of real and projected differences between results obtained
from simple, population-level models and more complex age- or size-specific models
that include behavior at the individual level proved the basis for arguments for
increased use of the latter. One part of the accumulation of new knowledge in science involves testing alternative explanations for observed phenomena. If all possible a priori explanations are not included, the inference scheme is limited.
More realistic models are needed to provide all possible explanations. For example,
age- and size-specific models led to a new possible explanation for depressed equilibria that had not yet been considered in any specific instance of this phenomenon.
Formulation of optimal policy can be viewed similarly. If all possible occurrences
(e.g., depressed equilibria, fishing causing instability) are not included in the
set of candidate policies, non-optimal policies may result (e.g., instability, constant rather than pulse fishing).

There are several specific criticisms of models as complex as those being
proposed here. One of these is that they require too much data. Shaeffer and
Beverton (1963) actually used the amount of data available as a criterion for choosing between the logistic and the more complex Beverton-Holt model. This issue can
be evaluated by discussing the actual gain of adopting each model. Plots of real
data in the form of yield versus effort for a logistic model increase from the
origin to a cloud of points that may or may not decrease as yield increases. This
may appear to be a "good fit" for two reasons: (a) at low levels yield increases
with effort, and (b) yield will eventually decrease as effort increases further.
Neither of these reflects much about population dynamics.

On the other hand, in use of more complex models one can follow the strategy
set forth by Sinko and Streifer (1969) for age-, size-specific models and MacFadyen
(1973) for ecological models in general. Complex models can be constructed based

on available data and reasonable assumptions where data are lacking, and can then
be used to determine behavior and optimal policy. Sensitivity of the results to
poorly known parameter values can then be determined and more data collected if
needed. This approach seems far better than adopting a less than realistic model
because of limited data. In either obtaining the necessary data or at least re-
alizing that not enough data exist to solve the problem, one avoids the complacency
to which logistic-type models can lead.

A second criticism of complex models is that they are too complex to understand. Rothschild and Suda (1977), in discussing models more complex than the
logistic, recently argued that "the complexity of these more complex models, while
still less complex than nature, soon outstrips comprehension and then at this point,
the complex model is, to the human mind, just as unfathomable as nature itself." I
disagree and would contend rather that attempts to understand more realistic models,
though they may be complex, will lead to increased understanding of the complex
processes in populations.

A third criticism is that mathematical analysis of complex models is too difficult, that is, not as many techniques are available for analysis of non-linear
partial differential equations as are available for simpler mathematical systems.
While this may be currently true, I don't think that enough attention has been paid
by mathematicians to use of realistic, complex models in solving real problems.
This is, in essence, my purpose in presenting the ideas in this paper. I hope to
elicit greater attention to these models in the future.

ACKNOWLEDGMENT

This work is a result of research sponsored by NOAA Office of Sea Grant,
Department of Commerce, under Grant #NOAA-M01-184 R/F52. The U.S. Government is
authorized to produce and distribute reprints for governmental purposes notwith-
standing any copyright notation that may appear hereon.

REFERENCES

Allen, R.L. and Basasibwaki, P. 1974. Properties of age structure models for fish
populations. J. Fish. Res. Board Can., 30, 1936-1947

Allen, K.R. and Kirkwood, G. 1977. A sperm whale population model based on cohorts
(SPCOH). Rep. Int. Waling Comm. (no. 27), 268-271

Beddington, J.R. and Taylor, D.B. 1973. Optimum age specific harvesting of a population. Biometrics, 29, 801-809

Beverton, R.J.H. and Holt, S.H. 1957. On the dynamics of exploited fish populations. Fish. Invest. London, Ser. 2, 19

Botsford, L.W. 1978. Modeling, stability and optimization of aquatic production systems. Ph.D. Thesis, University of California, Davis

Botsford, L.W. 1980a. The effects of increased individual growth rates on depressed population size. American Naturalist. In press

Botsford, L.W. 1980b. Optimal fishery policy for size-specific, density-dependent population models. Ms. submitted for publication

Botsford, L.W. and Wickham, D.E. 1978. Behavior of age-specific, density-dependent models and the northern California Dungeness crab fishery. J. Fish. Res. Board Can., 35, 833-845

Brodie, P.F. 1977. Form, function and energetics of Cetacea: a discussion. In Functional Anatomy of Marine Mammals, vol. 2, R.J. Harrison, ed. Academic Press, London

Clark, C. 1976. Mathematical Bioeconomics: The Optimal Management of Renewable Resources. John Wiley and Sons, Inc.

Clark, C., Edwards, G., and Friedlander, M. 1973. Beverton-Holt model of a commercial fishery: optimal dynamics. J. Fish. Res. Board Can., 30, 1629-1640

Getz, W.M. 1979. Optimal harvesting of structured populations. Math. Biosci., 44, 269-291

Graham, M. 1935. Modern theory of exploiting a fishery, and application to North Sea trawling. J. Cons. Expl. Mer, 10, 264-274

Guckenheimer, J., Oster, G. and Ipaktchi, A. 1977. The dynamics of density-dependent population models. J. Math. Biol., 4/2, 8-147

Gulland, J. 1977. The analysis of data and the development of models. In Fish Population Dynamics, J. Gulland, ed. John Wiley and Sons

Hannesson, R. 1975. Fishery dynamics: a North Atlantic cod fishery. Can. J. Econ., VIII(2), 151-173

Holling, C.S. 1973. Resilience and stability of ecological systems. Ann. Rev. Ecol. Syst., 4, 1-23

Hutchinson, G.E. 1978. An Introduction to Population Ecology. Yale University Press, New Haven and London, 260 pp.

Ivlev, V.S. 1961. Experimental Ecology of the Feeding of Fishes. Yale University Press, New Haven

Keyfitz, N. 1978. Introduction to the Mathematics of Populations. Addison-Wesley, Reading, 450 pp.

Laws, R.M. 1977. The significance of vertebrates in the Antarctic marine ecosystem. In Adaptations Within Antarctic Ecosystems, G.A. Llano, ed.

Leslie, P.H. 1945. On the use of matrices in certain population mathematics. Biometrika, 33(Pt. 3), 183-212

Levin, S.A. and Goodyear, C.P. 1980. Analysis of an age-structured fishery model. J. Math. Biology 9(3):245-274.

Lockyer, C.H. 1972. A Review of the Weights of Cetaceans with Estimates of the Growth and Energy Budgets of the Large Whales. M.Phil. thesis, University of London, London

Lockyer, C.H. 1974. Investigation of the ear plug of the southern sei whale, Balaenoptera borealis, as a valid means of determining age. J. Cons. int. Explor. Mer, 36(1), 71-81

Lotka, A.J. 1925. Elements of Physical Biology. Williams and Wilkins, Baltimore. Republished by Dover Publications, 1956, as Elements of Mathematical Biology

MacFadyen, A. 1973. Some thoughts on the behavior of ecologists. J. Anim. Ecol., 44, 351-363

MacKintosh, N.A. 1973. Distribution of postlarval krill in the Antarctic. Discovery Rep., 36, 95-156

May, R.M., Beddington, J.R., Horwood, J.W. and Shepherd, J.G. 1978. Exploiting natural populations in an uncertain world. Math. Biosci., 42, 219-252

May, R.M., Beddington, J.R., Clark, C.W., Holt, S.J. and Laws, R.M. 1979. Management of multispecies fisheries. Science, 205(4403), 267-277

Mendelssohn, R. 1976. Optimization problems associated with a Leslie matrix. The Amer. Natur., 110(973), 339-349

Menshutkin, V.V. 1964. Population dynamics studied by representing the population as a cybernetic system. (In Russian) Vopr. Ikhtiol., 4, 23-33

Murphy, G.I. 1968. Pattern in life history and the environment. The Amer. Natur., 102(927), 391-403

Paloheimo, J.E. and Dickie, L.M. 1965. Food and growth of fishes. I. A growth curve derived from experimental data. J. Fish. Res. Board Can., 22, 521-542

Pope, J.G. 1973. An investigation into the effects of variable rates of the exploitation of fishery resources. In The Mathematical Theory of the Dynamics of the Dynamics of Biological Populations, M.S. Bartlett and R.W. Hiorns, eds., Academic Press

Reed, W.J. In press. Optimum age-specific harvesting in a non-linear population model. Biometrics

Ricker, W.E. 1954. Stock and recruitment. J. Fish. Res. Board Can., 11, 559-623

Ricker, W.E. 1977. The historical development. In Fish Population Dynamics, J.A. Gulland, ed., John Wiley and Sons, New York, N.Y.,

Rorres, C. 1976b. Optimal sustainable yields of a renewable resource. Biometrics, 32, 945-948

Rorres, C. and Fair, W. 1975. Optimal harvesting policy for an age-specific population. Math. Biosci., 24, 31-47

Rothschild, B.J. and Suda, A. 1977. Population dynamics of tuna. In Fish Population Dynamics, J.A. Gulland, ed., John Wiley and Sons, London, 372 pp.

Salmon, W.C. 1973. Logic. Prentice-Hall, Inc., Englewood Cliffs, N.J., 150 pp.

Schaefer, M.B. 1954. Some aspects of the dynamics of populations important to the management of the commercial marine fisheries. Bull. Inter-Amer. Trop, Tuna Comm., 1, 26-56

Schaefer, M.B. and Beverton, R.J.H. 1963. Fishery dynamics - their analysis and interpretation. In The Sea, M.N. Hill, ed., 2, 464-483, Wiley, New York, N.Y.

Sinko, J.W. and Streifer, W.E. 1967. A new model for age, size structure of a population. Ecology, 48(6), 910-918

Sinko, J.W. and Streifer, W.E. 1969. Applying models incorporating age- size-structure of a population of Daphnia. Ecology, 50, 608-615

Von Foerster, H. 1959. Some remarks on changing populations. In The Kinetics of Cell Proliferation, F. Stohlman, ed., 382-407, Grune and Stratton, New York

Walters, C.J. 1969. A generalized computer simulation model for fish population studies. Trans. Amer. Fish. Soc., 98, 505-512

AGE-STRUCTURE AND STABILITY
IN MULTIPLE-AGE SPAWNING POPULATIONS

Simon A. Levin
Cornell University

Fishery models which are based upon stock-recruitment relationships in
which the total number of recruits decreases at large stock size (e.g.,
the Ricker model) are well-known to show very complicated dynamic behavior for some parameter values; in particular, the potential for the
population to maintain stable constant population size is strongly influenced by the growth rate at low densities, as measured by the parameter α, the density-independent egg replacement ratio. For multiple age-spawning populations, the situation is much more complicated:
increase in year to year survival for a given lifetime replacement
value α both spreads reproduction over more age-classes (which is
stabilizing) and introduces a delay into the system response (which
is potentially destabilizing) by increasing the mean age of reproduction. In this paper, the interplay of these two effects is explored.

INTRODUCTION

Much theory in fisheries science is based on the construction of stock-recruitment relationships under the assumption that individuals spawn once, and at
a common age. Most populations of interest do not fit this simplification, but
implicit in the application of the method is the hope that the assumption of a
dominant reproductive age will yield a good first approximation to population dynamics. For some species, this is reasonable; for others, such as the striped bass
(Morone saxatilis) reproduction may be spread over several age classes and it is
inadequate to treat population dynamics as if generations were non-overlapping.
Although the latter restriction may appear obvious, it has sometimes been ignored,
e.g., in attempts (McFadden 1977, McFadden and Lawler 1977) to assess the effects
of power plant entrainment and impingement mortality on Hudson River fish populations, including the striped bass.

Fisheries scientists have long been conscious of the problems of dealing with
multiple-age spawning. Ricker's classic 1954 paper "Stock and Recruitment" provides a profound treatment of the essential issues, and is still suggestive of new
insights a quarter-century later. Walters (1969) explored the effects by simulation models of age-structured populations, and in the last several years there has
been increased interest in the properties of such models (Allen and Basasibwaki
1974, DeAngelis et al. 1980, Silvert and Smith 1980, Travis et al. 1980). Other
studies (e.g., Clark 1976, Beddington 1977, Goh 1977), often motivated by investigations of whaling fisheries, have examined compromise models in which explicit

time lags are introduced. Basically, these break the population into two age classes - immatures and adults - by incorporating a non-trivial age of first reproduction. However, by treating all adults as reproductively identical, they ignore some of the subtle effects on age structure and population dynamics caused by more complicated fecundity schedules. In this paper, some of these effects will be explored.

TWO AGE CLASSES

The simplest model which explicitly incorporates age structure admits two age classes, with densities N_1 and N_2, with individuals dying after age two. (Excellent discussions of this case, from different perspectives, may be found in Guckenheimer et al. (1977) and in Pounder and Rogers (1980).) The probability of survival from age class 1 to age class 2 is denoted p . Further, associated with each age class j (=1, 2) there is a <u>maternity</u> m_j , the number of newborns in the next year spawned by an average individual of age j in a given year. Because density-dependent effects are most likely to have their impact upon the young-of-the-year, it is assumed that p is constant but that m_1 and m_2 may vary with density.

Following Goodyear (1975), define

K_j = average fecundity (number of eggs) of females of age class j,

and in any given year define the <u>parental egg production</u>

$$P = N_1 K_1 + N_2 K_2 \qquad (1)$$

It will be assumed that density-dependent effects occur mainly through competition among young-of-the-year, so that their force may be measured by P; more generally, were there inter-age class effects (as for example if there were cannibalism by older age classes upon the young), a different function would be needed to fully describe density-dependent effects (see Botsford and Wickham (1978), and Botsford, this volume). Restricting attention to the simpler case, assume

$$m_1 = \hat{\alpha} \hat{K}_1 F(P)$$
$$m_2 = \hat{\alpha} \hat{K}_2 F(P) \quad , \qquad (2)$$

where $\hat{\alpha}$ represents the density-independent survival and F(P) the density-dependent survival from egg to age 1.

For natural populations, the particular form of F(P) will not usually be known, although in general it is reasonable to stipulate that

$$F(0) = 1 \quad \text{and that} \quad F'(P) \leq 0 \quad . \qquad (3)$$

A number of particular forms for F(P) have been considered (May et al. 1978, Ludwig 1980), the most common being the <u>Ricker</u> relation

$$F(P) = \exp(-\beta P) , \qquad (4)$$

and the <u>Beverton-Holt</u> relation, which assumes $1/F$ to be linear in P . Often these and other hypothetical relationships are fitted to data; however rarely if ever can they be considered to be established. Errors in measurement of both stock and recruits are virtually unavoidable, and these create considerable statistical problems (Walters and Ludwig 1980, Ludwig and Walters 1980). Moreover, Ludwig (pers. comm.), in considering various candidate spawner recruit relations for 32 stocks of salmon (<u>Oncorhyncus</u> spp.), found the Ricker model to be inferior to a model of Cushing (1971), in which the number of recruits is proportional to a power of the number of spawners, in 12 of 14 cases where a choice could be made; the remaining 18 cases were inconclusive. Ludwig (1980) also performs similar analyses on populations of other species and again finds the Ricker relation to be something less than a universal truth. Despite these drawbacks, the Ricker curve remains the choice for many fisheries scientists, and hence it seems important to understand the properties of models based upon it. Thus although the problem will be cast in a general form the primary emphasis of this paper will be on Ricker models.

In the general case described above, the age-structured model takes the familiar form due to Leslie (1945, 1948)

$$N_1' = m_1 N_1 + m_2 N_2$$

$$N_2' = p N_1, \qquad (5)$$

in which N_1, N_2 represent densities in a given year and N_1', N_2' in the following year. Under the assumptions (2), these become

$$N_1' = \hat{\alpha} P F(P) ,$$

$$N_2' = p N_1 , \qquad (6)$$

in which $P = N_1 K_1 + N_2 K_2$.

The most familiar case of (6) is $F(P) \equiv 1$, in which there is really no density-dependence. As is well-known, in this case N_1 and N_2 will grow (or decay) asymptotically exponentially at a common rate. When the inequality in (3) is strengthened, so that $F'(P) < 0$, (6) admits a unique nontrivial equilibrium defined by

$$\bar{P} = F^{-1}(1/\alpha) , \qquad (7)$$

provided $\alpha = \hat{\alpha}(K_1 + pK_2) > 1$ and $\lim_{z \to \infty} F(z) < 1/\alpha$ (Note that $\bar{N}_1 = \frac{1}{p} \bar{N}_2 = \bar{P}/(K_1 + pK_2)$.) For the Ricker relation (4), (7) becomes

$$\bar{P} = (\ell n \ \alpha)/\beta \ . \tag{8}$$

The quantity $K_1 + pK_2$, which appears in the formulas for $\bar{N}_1$ and α , may be thought of as the <u>stock</u> <u>value</u> (V_s) of a one-year old; α is the stock value of an egg, that is the expected life-time replacement in eggs per egg, ignoring density-dependent effects.

The local stability of (7) may be studied by standard linearization techniques; the equilibrium is locally stable provided the eigenvalues of the matrix

$$M = \begin{bmatrix} K_1 Q & K_2 Q \\ p & 0 \end{bmatrix} , \tag{9}$$

are all less than 1 in magnitude, and unstable if any exceeds 1 . Here

$$Q = [1 + P \frac{d}{dP} \log F(P)]\big|_{P=\bar{P}} / (K_1 + pK_2) \ .$$

It is easily shown that the conditions for stability are that

$$-K_1 Q - 1 < -K_2 Qp < 1 \ . \tag{10}$$

In the case of the Ricker model, Q reduces to

$$Q = (1 - \ell n \ \alpha)/(K_1 + pK_2) \ ;$$

the stability conditions become

$$\ell n \ \alpha < 2 + K_1/K_2 p \ , \tag{11}$$

and if $p < K_1/K_2$,

$$\ell n \ \alpha < 2 + 2/(K_1/K_2 p - 1) \ . \tag{12}$$

If $K_1 \geq 2K_2$, then condition (12) is both necessary and sufficient; the stability region is as shown in Fig. 1, where the bottom axis is the mortality $z = -\ell n \ p$. For this case it is clear that decreasing mortality, which spreads reproduction more evenly over the two age classes, leads to increasing stability.

However, if $K_1 < 2K_2$, both conditions (11) and (12) are relevant, and the stability region has a more complicated general form shown in Figure 2 provided $K_1 > K_2$ and in Figure 3 provided $K_1 < K_2$ (that is, provided fecundity is larger in the older age class).

In these latter cases, the cusp-shaped peak of the stability region may seem surprising: it implies that decreasing mortality is stabilizing only up to a certain point, beyond which it may become destabilizing because it introduces a delay

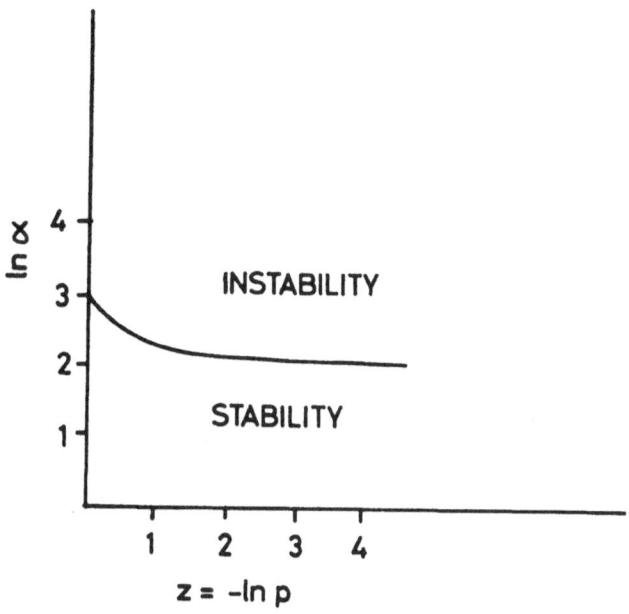

FIGURE 1. Stability diagram for two-age class model with $K_1 = 3K_2$

in the response of the system (see the more general discussion in Levin and
Goodyear 1980).

For a given p, as $\ln \alpha$ is increased beyond the critical value defined by
the curve in Fig. 1, or (for $p < K_1/2K_2$) as $\ln \alpha$ is increased beyond the values
defined by the right branches of the boundaries in Figs. 2 and 3, the equilibrium
solution defined by (8) becomes unstable by virtue of a real eigenvalue of M be-
coming less than -1. When this happens, a stable periodic solution of length 2
generations will bifurcate from (8), similar to those discussed in the single age-
class model by May (1974, 1978). Destabilization is in this case a characteristic
of the Ricker stock-recruitment relation and of others which similarly permit over-
shoot of the equilibrium (a Beverton-Holt model would not exhibit such behavior);
it has nothing to do with the age-structure of the population (although of course
the age structure will also oscillate). Thus in general increasing survival and
the consequent increased spreading of reproduction makes destabilization by over-
shoot less likely. However, it may increase the likelihood of destabilization by
another mechanism, discussed below.

The left branches of Figs. 2 and 3 have a somewhat different interpretation.
For points on them the determinant of M equals 1, so that (see (11)) its eigenvalues satisfy

$$\lambda^2 + u\lambda + 1 = 0 , \tag{13}$$

where $u = K_1/pK_2$. Note that along either left branch, $0 \le u \le 2$ and so (except at the peak) the defining relation $|\lambda| = 1$ implies that λ is complex; setting $\lambda = e^{i\theta}$ in (13) leads at once to the solution $\theta = \cos^{-1}(-u/2)$. However it is useful to obrain an approximation for θ which will illustrate a fairly general principle.

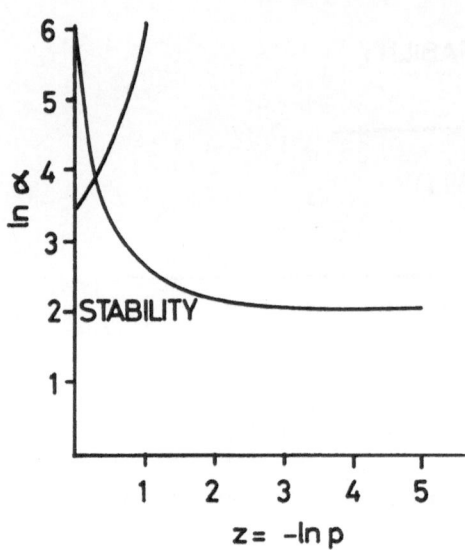

FIGURE 2. Stability diagram for two-age class model with $K_1 = 1.5K_2$

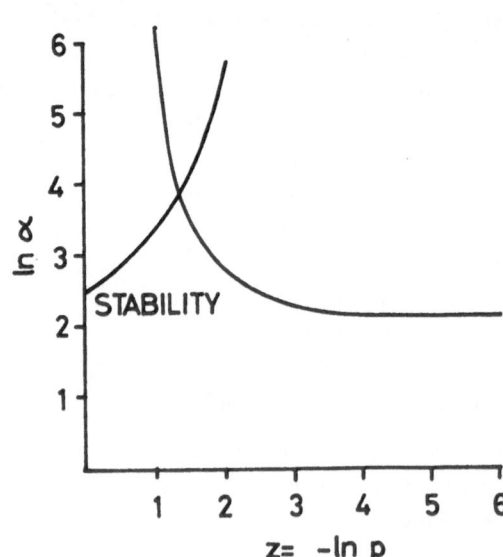

FIGURE 3. Stability diagram for two-age class model with $K_1 = .5K_2$

Define the mean age of reproduction, μ by

$$\mu = (K_1 + 2pK_2)/(K_1 + pK_2) = (u + 2)/(u + 1) \quad . \tag{14}$$

If generations were non-overlapping, the simplest bifurcation would lead to cycles twice the mean generation time; it is reasonable to look for similar phenomena in this case. Indeed, Ricker (1954) observed such bifurcations as a general phenomenon, although as I shall show the relationship is not as exact as Ricker implied (see also Keyfitz 1972, Lee 1974, Botsford and Wickham 1978). Using (14) to replace u in (13), and substituting $\lambda = e^{i\theta}$, one obtains

$$e^{2i\theta} + \frac{2-\mu}{\mu-1} e^{i\theta} + 1 = 0 \quad . \tag{15}$$

Adapting a method from demography of Frauenthal (1975; see also Botsford and Wickham 1978), multiply both sides of (15) by $e^{-i\mu\theta}$ to yield

$$e^{i\theta(2-\mu)} + \frac{2-\mu}{\mu-1} e^{i\theta(1-\mu)} + e^{-i\mu\theta} = 0 \quad . \tag{16}$$

The imaginary part of this is

$$\sin(2-\mu)\theta + \frac{2-\mu}{\mu-1} \sin(1-\mu)\theta = \sin \mu\theta \; ; \tag{17}$$

note that $0 < \theta \le \pi$, and that $\frac{4}{3} \le \mu \le 2$. Expanding $\sin(2-\mu)\theta$ and $\sin(1-\mu)\theta$ in power series, and ignoring all terms of third order or higher, one obtains the approximation

$$\sin \mu\theta \approx 0 \; , \tag{18}$$

which leads to the estimate $\mu\theta \approx \pi$. In general, there is no *a priori* justification for ignoring the higher order terms; inclusion of the cubic ones, for example, changes the estimate to

$$\sin \mu\theta \approx \frac{1}{3!} [(2-\mu)(3-2\mu)]\theta^3 \quad . \tag{19}$$

However, if $\mu > \sqrt{2}$, then from (15)

$$\theta = \cos^{-1}\{(\mu-2)/2(\mu-1)\} < 3\pi/4 \; ,$$

and from (19) it is easily shown that

$$|\sin \mu\theta| < 9\pi^3/1024 \approx .2725 \quad .$$

Thus for $\mu > \sqrt{2}$, $\mu\theta$ as estimated from (19) would vary by less than 9% from the value of π ; moreover the higher order terms are negligible. The remaining allowable values of μ lie in the narrow boundary region between 4/3 (1.333) and $\sqrt{2}$ (1.414); there the estimate $\mu\theta \approx \pi$ is not as good, but even in the worst case ($\mu = 4/3$) the actual value ($3\pi/4$) of $\mu\theta$ differs from the estimate π by only 25%. Thus, to a good first approximation, $\mu\theta \approx \pi$, and so $\lambda^{2\mu} \approx 1$. The bifurcation observed across the left-branch will be a Hopf-bifurcation (Guckenheimer et al. 1977; Wan 1978), and the bifurcating solution will have period $2\pi/\theta$. Therefore, the period of the bifurcating solution is approximately 2μ, that is, approximately twice the mean age of reproduction; this number must lie between 2 and 4. An eigenvalue λ will not usually be an exact root of unity. Therefore (Guckenheimer et al. 1977; see also Wan 1978 for the derivation of the stability condition for the Hopf bifurcation), usually the bifurcation will result in the emergence of a stable, small amplitude invariant curve about the fixed point $(\bar{N}_1, \bar{N}_2)$ in (N_1, N_2)-space. This means that solutions which are not quite periodic will arise, and with period of about 2μ.

If $\lambda^k = 1$ for small integer k, in general stable periodic solutions will emerge. However there are exceptions, (when k < 5) and the most striking occurs when the destabilizing eigenvalue is an exact cube root of unity. I thank John Guckenheimer for pointing out to me that Wan's demonstration that the Hopf bifurcation is in the "forward" direction does not apply to this special case, as is clear from condition (d) of Wan's introductory section. This situation occurs if $p = 1$ and $K_1 = K_2$ in (6), an example studied in great detail by Guckenheimer et al. (1977). Note that in this case, $\mu = 1.5$, and naively one expects a small-amplitude solution of period approximately 3 to emerge. Indeed what happens is that a solution of period exactly 3 arises, but it is unstable; however, there also exists a larger amplitude solution, also of period 3, which is stable. As Guckenheimer et al. (1977) demonstrate, destabilization of the equilibrium (8) does not occur until $\frac{\alpha}{2} = \hat{\hat{a}} = e^3/2 \approx 10.04$; however, when $\frac{\alpha}{2} = \hat{\hat{a}} \approx 8.95$, two large amplitude cycles of period 3 appear, one stable and the other unstable. For $8.95 < \hat{\hat{a}} < 10.04$, both the point equilibrium and one 3-cycle are stable, and the asymptotic behavior of a solution is dependent on initial conditions; this is a surprising phenomenon. As $\hat{\hat{a}}$ is reduced to 8.95 from above, the unstable 3-cycle and the stable one annihilate each other; as $\hat{\hat{a}}$ is increased to 10.04, the unstable periodic solution collapses to the equilibrium point. As I shall show in the next paragraph, it reemerges on the other side of 10.04 (Note that $\hat{\hat{a}} = K_1\hat{\alpha}$).

A simple calculation allows direct computation of both 3-cycles, and allows independent demonstration of their existence by methods somewhat different from those used by Guckenheimer et al. (1977). Denote the three successive values of N_1 on a 3-cycle by A, B, C; then it is easily seen from (6) that

$$A = \frac{\alpha}{2}(C + B) \ e^{-\beta(C+B)}$$

$$B = \frac{\alpha}{2}(A + C) \ e^{-\beta(A+C)}$$

$$C = \frac{\alpha}{2}(B + A) \ e^{-\beta(B+A)} \ . \tag{20}$$

In (N_1, N_2) space, these solutions have the repeating pattern (A, C), (B, A), (C, B); and by the symmetry of the problem as regards the two age classes, one is led to look for solutions in which one of these three pairs lies on the line $N_1 = N_2$. Thus, without loss of generality, assume B = C, so that (20) reduces to

$$A = \frac{\alpha}{2}(2B) \ e^{-\beta 2B}$$

$$B = \frac{\alpha}{2}(A + B) \ e^{-\beta(A+B)} \ .$$

Define

$$x = (A+B)/2B \ .$$

Then clearly $\frac{1}{2} < x < \infty$, and $x \neq 1$ except when the three-cycle degenerates to a point. Moreover, from (20), it is easily shown (for $x \neq 1$) that for any 3-cycle

$$h(x) = (2x-1)^{\frac{x}{x-1}} \ x^{\frac{1}{x-1}} = \alpha = 2\hat{\alpha} \ , \tag{21}$$

and conversely that (assuming $x \neq 1$) every value of $x > \frac{1}{2}$ satisfying (21) allows definition of a unique positive 3-cycle for which the pattern of N_1- values is A, B, B, ... The graph of $h(x)$ is shown in Fig. 4. It decreases from ∞ monotonically as x is increased until a unique minimum is reached at about $x = 1.73$, at which point $h(x) \approx 17.888$; beyond this point, $h(x)$ rises monotonically to ∞ as x is increased without bound. Note that although strictly speaking $h(x)$ is not defined at $x = 1$, its definition may be extended smoothly to include $x = 1$.

As is clear from Fig. 4, to every value of $\alpha > 17.888$ - that is, to every value of $\frac{\alpha}{2} = \hat{\alpha} > 8.944$ - there exist two 3-cycles, one of which degenerates to a fixed point when $\frac{\alpha}{2} = \hat{\alpha} = \frac{e^3}{2} \approx 10.04$. One can view x as a measure of the amplitude of the cycle, since $x = \frac{1}{2} + \frac{1}{2} \frac{A}{B}$, where the N_1-values in the cycle have the pattern A, B, B, ... Thus, when x is large, the cycle is characterized by a high value followed by a repeated low value; whereas if $x < 1$, the reverse situation obtains. For $\hat{\alpha} < 8.944$, no 3-cycles exist, but a pair of high-low-low ones emerge as $\hat{\alpha}$ passes this critical value; at the critical value of the parameter, these will be identical and neutrally stable. As $\hat{\alpha}$ is increased beyond the critical value, the two 3-cycles separate, and it may be shown numerically that the one with the larger x value is stable. As $\hat{\alpha}$ is increased through 10.04, the unstable 3-cycle shrinks to a point, merging at exactly $e^3/2$ with the equilibrium point given by (8), which is stable until then. Beyond $e^3/2$, the equilibrium point is unstable, and the unstable 3-cycle reemerges as an unstable low-high-high cycle. Once x is determined from Fig. 4, it is trivial to compute the values of A and B from (21).

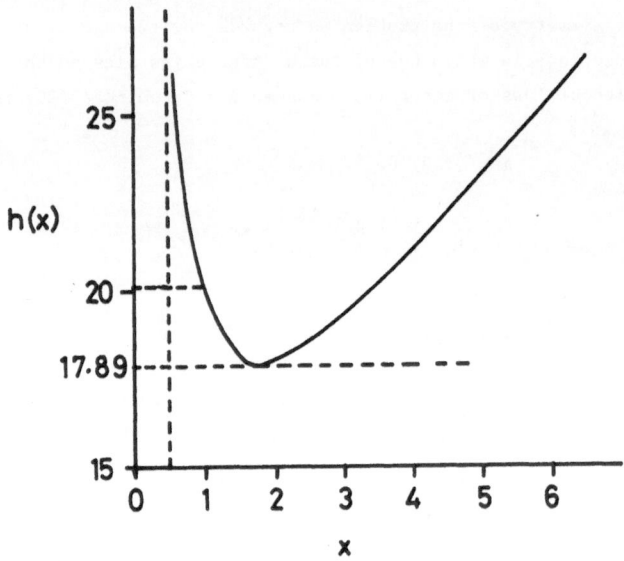

FIGURE 4. Graph of $h(x) = (2x-1)^{\frac{x}{x-1}} x^{\frac{1}{x-1}}$

This case is a somewhat special one; but as Guckenheimer et al. (1977) discuss, even when the destabilizing eigenvalue is not a cube root of unity (e.g., when $K_1 \neq K_2$), a large amplitude periodic solution will still coexist with the stable fixed point for some range of values of α . In this case, however, as α is increased beyond a critical value a stable small amplitude in variant circle emerges and coexists with the stable periodic solution for some range of the parameters. As α is increased still further, a complicated sequence of bifurcations will result, eventually culminating in a chaotic region (Guckenheimer et al. 1977).

Although the two age class case is obviously only the simplest possible model with age structure, it illustrates most of the phenomena observed in more complicated models for the kinds of fecundity schedules which one generally observes. This suggests that a two age class approximation may represent a reasonable compromise approach to the modelling of many more complicated age-structured models involving multiple-age-spawning; this idea is however only a preliminary one, and requires further investigation.

In summary, what is observed is that if reproduction is concentrated in the first age-class (specifically if $K_1 > 2K_2$), then increasing year-to-year survival p is stabilizing in that increasing p increases the range of α -values consistent with stability. In this case, for any p there is a critical value of α beyond which oscillations of period 2 arise due to the "over-shoot" property of the underlying stock-recruit relation. If, on the other hand, $K_1 < 2K_2$, then the stabilization by increasing p applies only until $p = K_1/2K_2$. Beyond that point, there is a transfer in the spectrum of the linearization matrix from a real eigenvalue to a pair of complex ones, and increasing p further is destabilizing because it increases the time-lag of response of the system. As p increases, the mean age of reproduction also increases; and for given p, as α is increased beyond a critical value (which decreases with p), a Hopf bifurcation takes place giving rise to a solution with period approximately equal to twice the mean age of reproduction. Generally, even before destabilization of the fixed point occurs, the period of the eventual bifurcating solution can be observed in the damped oscillations to equilibrium. As α is increased further, more complicated behavior results. The overall picture is summarized in Figs. 1-3; in Figs. 2-3 bifurcation to the left of the peak may be thought of as being induced by age-structured effects; as p increases, there are more age classes and the system thus takes longer to respond to perturbations. Again, instability is related to the choice of stock-recruit relationship: for example, the Beverton-Holt model will not exhibit instability.

A peculiar property of the single age-spawning Ricker model is that periodic solutions demonstrate population averages independent of the length of the period, and thus identical to the value of the equilibrium solution; however this property is lost for multiple age-spawning populations. This is an extremely important practical point, for it means that for example the methods utilized by McFadden (1977), McFadden and Lawler (1977) of computing reduced "equilibrium" population sizes of the striped bass would not give reliable information concerning average behavior if population sizes fluctuate due to α values too high.

The behavior of the two age class model may become easier to understand if (6) is rewritten as a second-order difference equation. For this purpose, define the function

$$G(P) = \alpha PF(P) \quad . \tag{22}$$

Then (6) leads to the relation

$$P(t+2) = \frac{K_1}{K_1+pK_2} G[P(t+1)] + \frac{pK_2}{K_1+pK_2} G[P(t)] \quad , \tag{23}$$

where $P(t)$ is the parental egg production in year t. If $p = 0$ this relationship

degenerates to the single age spawning model, and for p small the first term on the
right of (23) dominates. For larger p, the larger (two-generation) time lag also
becomes important, introducing an oscillation into the dynamics. Related discussions may be found in Keyfitz (1972) and Lee (1974).

Essentially the same picture emerges even if $K_1 = K_2$: Fig. 3 applies. In
that case, it is easily seen that the left branch of the stability curve owes its
existence to the <u>truncation</u> of the population beyond age 2. If instead it were
simply assumed that all individuals had equal fecundity, and <u>p</u> were the probability
of year-to-year survival independent of age, then (6) could be replaced by the
single equation

$$P' = \hat{\hat{\alpha}} P F(P) + pP \quad . \tag{24}$$

As before, the equilibrium is defined by (7), where now

$$\alpha = \hat{\hat{\alpha}}/(1 - p) \quad ;$$

in the case of the Ricker model, the equilibrium is easily shown to be stable
provided

$$\ln \alpha < 2/(1 - p) \quad . \tag{25}$$

The effect of increasing p is unambiguous, it is stabilizing: the larger the p ,
the larger the range of α values consistent with stability.

A delay in reproduction may be introduced, still without truncation, by taking
$K_1 = 0$ and $K_i = K$ thereafter. Then, in place of (24), one obtains

$$P(t+2) = pP(t+1) + p\hat{\hat{\alpha}}P(t)F(P(t)) \quad . \tag{26}$$

In this case,

$$\alpha = \hat{\hat{\alpha}}p/(1 - p) \quad .$$

Assuming again the Ricker form, the equilibrium (7) is stable provided

$$\ln \alpha < 1 + \frac{1}{1-p} \quad ; \tag{27}$$

Once again, the larger the p, the larger the range of $\ln \alpha$ consistent with stability: the absence of the truncation effect means that increasing p is always stabilizing. I shall return to this theme in the next section.

It is worth pointing out that if (27) were rephrased in terms of $\hat{\hat{\alpha}}$, the resulting inequality

$$\ln \hat{\hat{\alpha}} < \ln((1 - p)/p) + 1 + 1/(1 - p) \tag{28}$$

would lead to somewhat less transparent interpretation: the right hand side has a
minimum for p = 0.5 . This complication is because (28) reflects not only the ef-

fects underlying (27), but also the fact that varying p while keeping $\hat{\hat{\alpha}}$ fixed affects the growth potential (as reflected by α) of the population. More complicated versions of (26), with longer reproductive delays or with other stock-recruit relationships of the same general form, show essentially the same stability behavior (Clark 1976, Goh 1977, Goh and Agnew 1978, Beddington 1977, Beddington 1980). It therefore seems preferable to couch the discussion of stability in terms of α and p (rather than $\hat{\hat{\alpha}}$ and p) so that the two separate influences of alteration in p are not confounded. If this is done, the anomalous behavior of the models of Clark (1976) and Goh (1977) is avoided.

MORE THAN TWO AGE-CLASSES

The approach described in the previous section can be easily extended to an arbitrary number of age classes. Following Levin and Goodyear (1980) and Goodyear (1980), let N_j be the number of females in age class j, m_j the maternity associated with that age class, and p_j the probability of survival from age j to age $j + 1$. p_j is assumed independent of density.

Define

K_j = average fecundity (number of eggs) of females of age class j, and in any given year, the parental egg production

$$P = \sum_j N_j K_j \quad .$$

As before

$$m_j = \hat{\alpha} K_j F(P) \quad . \tag{29}$$

where

$\hat{\alpha}$ = density independent probability of surviving from egg to age 1. Note that the net maternity function is determined up to a constant multiple by the fecundities K_j.

The dynamics may then be completely described by the system

$$N' = MN \quad , \tag{30}$$

where

$$N = \begin{bmatrix} N_1 \\ N_2 \\ \cdot \\ \cdot \\ \cdot \end{bmatrix} \tag{31}$$

and

$$M = \begin{bmatrix} m_1 & m_2 & \cdot & \cdot & \cdot \\ P_1 & 0 & \cdot & \cdot & \cdot \\ 0 & P_2 & \cdot & \cdot & \cdot \\ \cdot & \cdot & \cdot & \cdot & \cdot \\ \cdot & \cdot & \cdot & \cdot & \cdot \\ \cdot & \cdot & \cdot & \cdot & \cdot \end{bmatrix} \qquad (32)$$

This system will have a point equilibrium defined by

$$\bar{N}_j = \ell_j \bar{N}_1 = P_1 \cdots P_{j-1} \bar{N}_1 \qquad (33)$$

$$\bar{p} = F^{-1}(1/\alpha)$$

provided the conditions (3) are met and

$$\alpha = \hat{\alpha} V_s = \hat{\alpha} \sum_j \ell_j K_j > 1 \quad . \qquad (34)$$

Note that the <u>survival probability</u> ℓ_j from age 1 to age j and the <u>stock value</u> V_s of an age 1 recruit are introduced in (33) and (34); by convention $\ell_1 = 1$.

By standard linearization methods, it is shown that the stability of the equilibrium is determined by the roots of the equation

$$1 = w_1 \ell_1 \lambda^{-1} + w_2 \ell_2 \lambda^{-2} + \ldots , \qquad (35)$$

where

$$w_j = K_j \frac{\partial}{\partial P} (\hat{\alpha} P F(P))\big|_{N=\bar{N}} = \frac{K_j}{V_s} (1 + \bar{P} F'(\bar{P})/F(\bar{P})) \qquad (36)$$

(35) may be rewritten

$$\frac{1}{1 - \gamma} = \sum_j \ell_j f_j \lambda^{-j} , \qquad (37)$$

in which

$$f_j = K_j / \sum_j \ell_j K_j \qquad (38)$$

and

$$\gamma = - \bar{P} F'(\bar{P})/F(\bar{P}) > 0 \quad . \qquad (39)$$

(In the case of the Ricker model, $\gamma = \ln \alpha$.) The equilibrium will be stable to small perturbations if all roots of (37) are less than 1 in magnitude, and unstable if any exceeds 1 . Following the methods introduced for continuous populations by Frauenthal (1975), define the mean and the variance of the net maternity function by

$$\mu = \sum_j j \ell_j f_j \tag{40}$$

$$\sigma^2 = \sum_j (j-\mu)^2 \ell_j f_j \quad . \tag{41}$$

Note that μ is the mean age of reproduction.

Depending on the particular form of F, destabilization of the equilibrium may or may not occur as γ is increased for p fixed. However, if destabilization does occur the critical value $\gamma*$ of γ and the "period" of the bifurcating solution may be approximated by an adaptation of the method of Frauenthal. The assumption $\gamma > 0$ removes the possibility of $\lambda = 1$ as a bifurcating eigenvalue, but $\lambda = -1$ is a possibility. However, the boundary $\lambda = -1$ is easily found from (37); the region of stability is

$$\gamma \leq 1 + \cfrac{1}{\sum\limits_{j \ odd} \ell_j f_j - \sum\limits_{j \ even} \ell_j f_j} \quad . \tag{42}$$

If first bifurcation from the fixed point is via $\lambda = -1$, solutions of period 2 will emerge; thus the boundary (42) must be compared with that obtained by assuming that the destabilizing eigenvalue is complex. In applications, the "pitchfork" $\lambda = -1$ bifurcations occur in some cases and do not in others. In the case of the striped bass population (to be discussed later), the initial bifurcation is always by a complex eigenvalue, but period-doubling ($\lambda = -1$) bifurcations eventually occur as γ is increased further; these will be preceded by a sequence of other bifurcations.

If the destabilizing eigenvalue is complex, then (following Frauenthal 1975) set $\lambda = e^{i\theta}$ and multiply both sides of (37) by $e^{i\mu\theta}$ to yield

$$\frac{e^{i\mu\theta}}{1-\gamma*} = \sum_j e^{i(\mu-j)\theta} \ell_j f_j \quad . \tag{43}$$

Expand the exponentials on the right in a Taylor series about μ, and separate real and imaginary parts to obtain the asymptotic series

$$\frac{\sin \mu\theta}{1 - \gamma*} = \theta \sum_j (\mu-j) \ell_j f_j + \text{terms of third order} \tag{44a}$$

$$\frac{\cos \mu\theta}{1 - \gamma*} = \sum_j (1 - \frac{(\mu-j)^2 \theta^2}{2}) \ell_j f_j + \text{terms of fourth order} \ . \tag{44b}$$

Using (38), (40), and (41), and ignoring terms of higher order (an act of faith), one is led to the approximations

$$\mu\theta \simeq \pi \tag{45}$$

and

$$\frac{1}{\gamma^* - 1} \simeq 1 - \frac{\theta^2\sigma^2}{2} \quad . \tag{46}$$

Thus, as before, the period of the bifurcating solution will be approximately 2μ , and the value of γ^* will be given approximately by

$$\gamma^* \simeq 1 + \frac{1}{1 - \frac{\theta^2\sigma^2}{2}} \simeq 1 + \frac{1}{1 - \frac{\pi^2\sigma^2}{2\mu^2}} \quad ; \tag{47}$$

this should be compared with (42). Indeed, there is nothing about the above development that excludes $\lambda = e^{i\pi} = -1$. For (45) to remain valid in this case, $\mu \simeq 1$; that is, 2-cycles should only arise when most reproduction is confined to the first year. This is consistent with what is observed in a variety of models. (47) is also valid in this case, but obviously (42) is more accurate.

The approximation (47) is a very crude one: by ignoring the fourth moment, it consistently overestimates γ^* . It will be most accurate when the net maternity function has a tight distribution, but is subject to some error otherwise. Nonetheless, it can work surprisingly well in practice, and allows additional insights into the mechanisms affecting stability. An example follows.

Goodyear (1980) and Levin and Goodyear (1980) consider in great detail the dynamics of Ricker-type models for the Hudson River striped bass population, using fecundities taken from McFadden and Lawler (1977) and taking $p_i = p$ for all i (a realistic assumption).

By computer simulations (see for example Figure 5) the boundary of the stability region is delineated, and this is shown in Figure 6 in comparison with the (higher) approximation defined by (47). The two curves shown differ, but the general patterns are the same and show peaks at virtually the same value of $z = \ell n\, p$. For p small (z large), the agreement is excellent, because σ^2 is small relative to μ^2. Note that $\mu \geq 3$, since there is no reproduction in the first two years of life; therefore, bifurcation to 2-cycles never occurs. As p is increased, reproduction is spread (σ^2 increases), but because of the truncation of reproduction after age 15, σ^2 is bounded. Therefore, the benefits of increasing p to spread reproduction are limited. On the other hand, as p is increased, μ continues to increase and the ratio σ^2/μ^2 decreases. Thus, (47) predicts the observed peak to the stability region at an intermediate p, and to a good approximation this occurs when σ/μ achieves its maximum. (With the fecundity schedule given in Table 1 the second term in (47) will be positive for all p.)

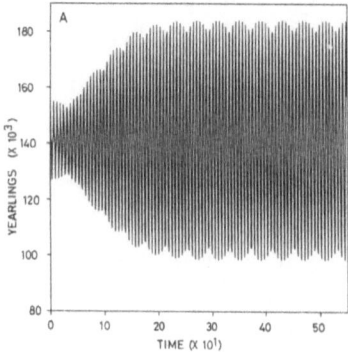

FIGURE 5. Simulation of the striped bass population, for $\ln\alpha = 2.10$, $z = 3$, showing number of yearlings versus time. Values shown are to be multiplied by numbers shown in parentheses (from Levin and Goodyear, 1980)

With Mark Kot, I have studied the nonlinear behavior of the striped bass model in more detail. When destabilization occurs, the suggestion of Ricker and of the preceding analysis that an almost periodic solution of period about twice the mean age of reproduction will bifurcate proves to be a very good one, (see e.g., Fig 7). As α is increased further beyond the critical value, a sequence of bifurcations occur, similar to those observed by Guckenheimer, et al. (1977) for their model. Some of these are through an eigenvalue +1, increasing the period of the oscillation by 1; others are period-doubling "pitchfork" bifurcations through -1; others will be of a more complicated nature.

The general pattern of the stability boundary follows that observed in the two age class case. The effect at low p is accentuated by the existence of an explicit reproductive delay, but all that is necessary is that for increasing age there be a rapid rise in fecundity in the early age classes. Similarly, the explicit truncation after 15 age classes makes crisper the fall-off in the stability boundary for large p, but all that is necessary is that fecundity level off in the later age classes and then eventually decline (if the latter condition were not met, the stability boundary would rise a second time for very high p). Thus the general lessons of the two age class case seem to remain valid for a wide class of reasonable fecundity schedules.

It should be noted that for the Beverton-Holt model, γ is always less than 1, and so instability is not possible for any p .

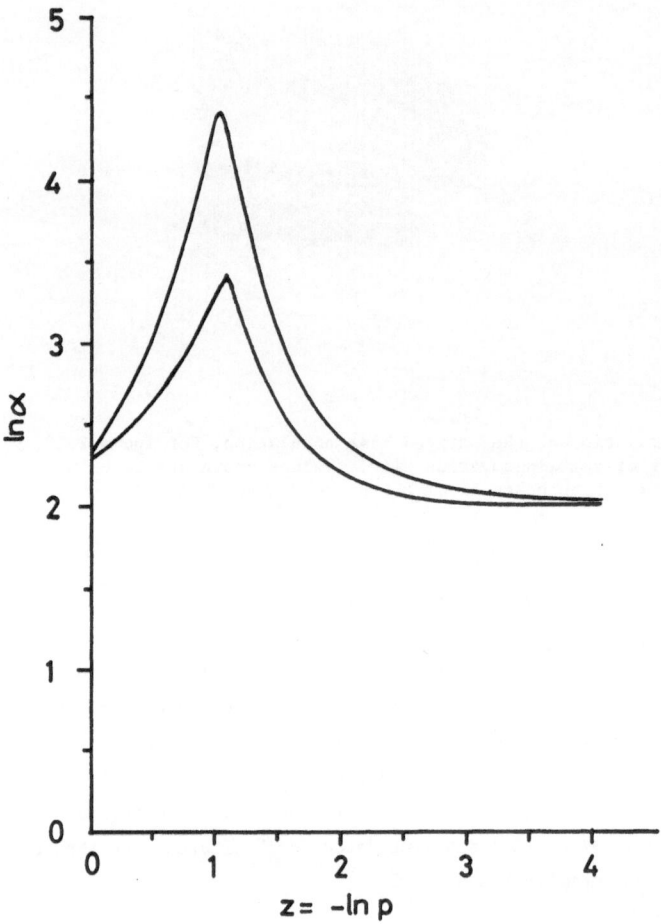

FIGURE 6. Comparison of actual (lower) and approximated (upper) stability boundaries for the striped bass model.

CONTINUOUS MODELS

Although the focus of this paper is on discrete time models, analogous behavior results if one considers time as continuous. In this case, one describes the age structure of populations at time t by the density function n(a,t), where a denotes age. Define the rate of recruitment

$$R(t) = \hat{\alpha}B(t)F(B(t)) , \qquad (48)$$

where F is as in (3) and B(t) is the rate of reproduction

$$B(t) = \int_0^\infty n(a,t)K(a)da \quad . \tag{49}$$

Here K(a) is the rate of production of eggs per individual of age $\underline{a}$. Finally, note that

$$n(a,t) = n(0, t-a) \ \ell(a) = R(t-a) \ \ell(a) \tag{50}$$

where $\ell(a)$ is the survival fraction from age 0 to age $\underline{a}$. Combining these equations yields the renewal equation

$$B(t) = \int_0^\infty \hat{\alpha}B(t-a)F(B(t-a)) \ \ell(a)K(a)da \quad . \tag{51}$$

This equation clearly admits the constant solution

$$\bar{B} = F^{-1}(1/\alpha) \ , \tag{52}$$

provided

$$\alpha = \int_0^\infty \hat{\alpha} \ \ell(a)K(a)da > 1 \quad . \tag{53}$$

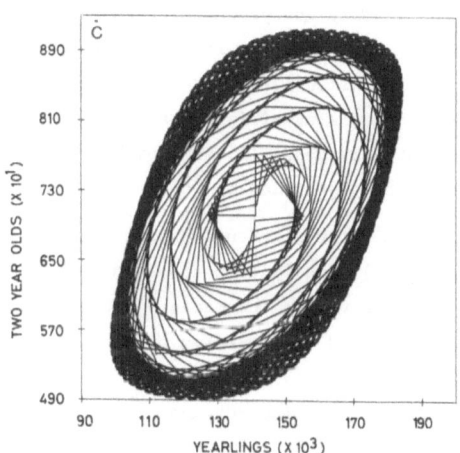

FIGURE 7. Simulation of the striped bass population, for $\ln\alpha = 2.10$, $z = 3$, showing number of two year olds versus number of yearlings. Values shown are to be multiplied by numbers shown in parentheses (from Levin and Goodyear, 1980).

The equilibrium is stable to small perturbations provided the roots of

$$\frac{1}{1-\gamma} = \int_0^\infty \ell(a)K(a)e^{-\sigma a}da \left/ \int_0^\infty \ell(a)K(a)da \right. \tag{54}$$

all have negative real parts. Here

$$\gamma = - \bar{B}F'(\bar{B})/F(\bar{B}) > 0 \quad . \tag{55}$$

From this point on, the analysis proceeds as before; indeed the problem is now in exactly the form for which Frauenthal developed his methods, and so it may be easily shown that (45) and (47) apply, with μ and σ defined appropriately (Frauenthal, 1975; Botsford and Wickham, 1978). As before, solutions of approximate period 2μ bifurcate when the constant solution becomes unstable. I leave a further discussion of this case for a future paper.

CONCLUSIONS: MANAGEMENT IMPLICATIONS

Fishery models which are based upon stock-recruitment relationships in which the total number of recruits decreases at large stock size (e.g., the Ricker model) are well-known to show very complicated dynamic behavior for some parameter values; in particular, the potential for a population to maintain stable constant population size is strongly influenced by the growth rate at low densities, as measured by the parameter α, the density-independent egg replacement ratio. For multiple age-spawning populations, the situation is much more complicated: increasing year to year survival for a given lifetime replacement value α both spreads reproduction over more age-classes (which is stabilizing) and introduces a delay into the system response (which is potentially destabilizing) by increasing the mean age of reproduction. In this paper, the interplay of these two effects has been explored.

In particular, if one focuses attention upon α and the year-to-year survival p (taken independent of age), one finds that maximum stability occurs at intermediate levels of p . For a given p, the maximum value of α consistent with stability is approximately determined by the ratio of the standard deviation to the mean of the net maternity function, with larger values of this ratio corresponding to an increased "stable" range for α . The survival probability p* which results in maximal stability is approximately the value of p which maximizes the standard deviation to mean ratio. In general, when instability results, a Hopf bifurcation occurs and a stable quasiperiodic solution of period approximately twice the mean age of reproduction will bifurcate from the constant solution. In most cases, this would be expected to be of small amplitude, but there are cases when it is of large amplitude and may actually make its appearance before the constant solution becomes unstable. In these cases, a small amplitude solution still emerges at the point of instability, but it is unstable.

These results have a number of potentially important implications for management. Further, as pointed out by Beddington and May (1977), even if deterministic models show populations to be stable at constant levels, it is important to understand "how stable," as defined by the dominant eigenvalue of the linearization, in order to understand how well buffered the population will be to random perturbations. Thus linearization analysis of the sort indicated in this paper is informative even within the region of stability.

A normal fishery would be expected to affect the survival values p_j, for j equal to or larger than the age of first fishing; power plants, on the other hand, if their major influences were the entrainment and impingement of young-of-the-year, would have principal effect upon the density-independent survival $\hat{\alpha}$. Thus it is clear that power plants will not operate like conventional fisheries, and that the two sources of impacts can have potentially very different effects upon multiple-age spawning populations. Both will reduce α and, in general, average population size, but their effects upon patterns of fluctuation of populations could be quite different. There is no simple way to define equivalencies between particular levels of fishing and specified power plant impacts.

Before discussing these notions further, it is important to note that no easy answer is available as to whether fluctuations are "good" or "bad" for populations. Natural populations can exhibit substantial fluctuations, to some extent generated by density-independent factors and to some extent related to density-dependent causes such as those discussed in this paper. Any theories advanced to suggest that fluctuating populations are either better able or less able to deal with environmental variation are almost always loaded down with group-selective arguments. From the viewpoint of the manager, a widely fluctuating population is obviously less predictable as a resource than one with more tightly controlled fluctuations. However whether it is more or less likely to crash is unclear, especially when the fluctuating population has a higher mean density.

In the simplest kind of fishery, in which all potentially reproductive individuals are subjected to the same fishing pressure, the effects of fishing can be evaluated by varying p while holding α constant. In this case, it would perhaps be preferable to revert to the formation of Clark (1976) and Goh (1977), where the stability region is defined in terms of p and $\hat{\alpha}$ rather than p and α. However, results can be obtained as well by superimposing curves of constant $\hat{\alpha}$ on (z, α) stability diagrams, as in Fig. 8. For the example shown, based on the Ricker model for the striped bass, decreasing p (increasing mortality z) while holding $\hat{\alpha}$ fixed reduces population fluctuations; however, in other situations (Goh 1977, Goh and Agnew 1978), it could be "stabilizing" for part of the range of p and destabilizing for others.

More usually, fishing would be concentrated on older age classes; further, density-dependent effects may involve inter-age class interactions. Under these conditions, Botsford and Wickham (1978) have shown that fishing pressure can be destabilizing as well as stabilizing. A gratuitous observation is that fishing of

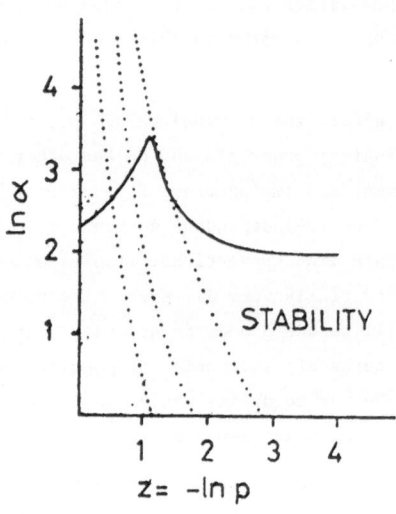

FIGURE 8. Stability diagram for
the striped bass model, with curves
of constant $\hat{\alpha}$ superimposed.

Curves shown are from left,
respectively $\ln \hat{\alpha} = -9, -7, -5$.

oldet age classes is to some extent a truncation of the older classes, as discussed in this paper, and this may provide some explanation of Botsford and
Wickham's observations. The general lesson seems to be that anything which reduces
the ratio of the standard deviation to the mean of the net maternity function is
potentially destabilizing. Moreover, by reducing the mean age of reproduction,
selective fishing of older age classes will shorten the period of fluctuation of
the population, and this effect will be observed even in the damped oscillations
within the stable region.

A power plant impact would operate somewhat differently, principally affecting
$\hat{\alpha}$ and leaving p unchanged, unless there were direct or indirect effects upon older
individuals. Reducing $\hat{\alpha}$ while leaving p unchanged is equivalent to reducing α
while leaving p unchanged, and in general this will tend to damp fluctuations but to
leave invariant the period of fluctuation. However, within the unstable region,
the nonlinear dynamics can be very complicated, and even within the stable region
the possibility of multiple stable states must be considered. Hence facile answers
are to be avoided. Moreover, it must be emphasized that the observed dynamics are
likely to be strongly influenced by the choice of stock-recruitment relationship;
the difficulties in making this choice have already been discussed. Too often insufficient attention is directed to this point, and predictions made too cavalierly
must be highly suspect.

As stated earlier, no judgment is suggested as to whether fluctuating populations are better off or worse off than more constant ones. However, any manipulation which will vary the natural pattern of variation of a population will have a potentially important impact on that population; this comment applies whether the manipulation increases or decreases fluctuation. The analyses presented in this paper concentrate on how patterns of fluctuation will be affected when parameters are varied within some particular and simplified model systems. The ideas presented seem to demand deeper investigation. Of particular interest and importance, would be an extension to consideration of populations which are strongly influenced by density-independent factors. Such populations will not in general yield to such clean mathematical analysis; but since density-independent influences seem to be inescapable in nature, it is essential to include their effects in realistic attempts to develop strategies for management and control.

ACKNOWLEDGEMENTS

I am grateful for many useful conversations with my student Mark Kot, who performed most of the simulations on which my insights are based; Figures 5 and 7 are due to his efforts. I also happily acknowledge the National Science Foundation of the United States for its support under grant MCS 7701076, and to the John Simon Guggenheim Memorial Foundation and the University of British Columbia for their help and support during my sabbatical visit there.

REFERENCES

Allen, R.L. and Basasibwaki, P. 1974. Properties of age structure models for fish populations. J. Fish. Res. Bd. Can. 31, 1119-1125

Beddington, J.R. 1977. On the dynamics of Sei whales under exploitation. Report of the International Whaling Commission 28, 169-192

Beddington, J.R. 1980. Harvesting and population dynamics. Chapter 14. In: Population dynamics. Symposium of the British Ecological Society. (R.M. Anderson, B.C. Turner, L.R. Taylor, eds.) Blackwell.

Beddington, J.R. and May, R.M. 1977. Harvesting natural populations in a randomly fluctuating environment. Science 197, 463-5

Botsford, L.W. and Wickham, D.E. 1978. The behavior of age-specific, density-dependent models and the Northern California Dungeness Crab Fishery. J. Fish. Res. Bd. Can. 35, 833-843

Clark, C.W. 1976. A delayed-recruitment model of population dynamics with an application to baleen whale populations. J. Math. Biol. 3, 381-391

Cushing, D.H. 1971. The dependence of recruitment on parent stock in different groups of fishes. J. Cons. Int. Explor. Mer. 33, 340-362

DeAngelis, D.L., Svoboda, L.J., Christensen, S.W., and Vaughan, D.S. 1980. Ecol. Modelling. Stability and return times of Leslie matrices with density-dependent survival: applications to fish populations. (In press)

Frauenthal, J.C. 1975. A dynamic model for human population growth. Theor. Pop. Biol. 8, 64-73

Goh, B.S. 1977. Stability in a stock-recruitment model of an exploited fishery. Math. Biosci. 33, 359-372

Goh, B.S. and Agnew, T.T. 1978. Stability in a harvested population with delayed recruitment. Math. Biosci. 42, 187-197

Goodyear, C.P. 1980. Implications of population structure on the oscillatory behavior of a population controlled by a Ricker function. (In press)

Guckenheimer, J., Oster, G., and Ipatchi, A. 1977. The dynamics of density dependent population models. J. Math. Biol. 4, 101-149

Keyfitz, N. 1972. Population waves. pp. 1-35 in T.H.E. Greville (Ed.) Population dynamics. Academic Press, Inc. New York, N.Y.

Lee, R. 1974. The formal dynamics of controlled populations and the echo, the boom and the bust. Demography 11: 563-585

Leslie, P.H. 1945. On the uses of matrices in certain population mathematics. Biometrika 33: 183-212

Leslie, P.H. 1948. Some further notes on the use of matrices in population mathematics. Biometrics. 35: 213-245

Levin, S.A. and Goodyear, C.P. 1980. Analysis of an age-structured fishery model. J. Math. Biol. 9. (In press)

Ludwig, D. 1980. Management of fish stocks using catch and effort data. (ms).

Ludwig, D. and Walters, C. 1980. Measurement errors and uncertainty in parameter estimates for stock and recruitment. Can. J. Fish. Oceans. (In press)

May, R.H. 1974. Biological populations with nonoverlapping generations: stable points, stable cycles, and chaos. Science 186: 645-647

May, R.M. 1978. Mathematical aspects of the dynamics of animal populations. In: Studies in mathematical biology II: Populations and communities (S.A. Levin, ed.), pp. 317-366. Washington, D.C.: Mathematical Association of America

May, R.M., Beddington, J.R., Horwood, J.W., and Shepherd, J.G. 1978. Exploiting natural populations in an uncertain world. Math. Biosci. 42: 219-252

McFadden, J.T. 1977. Influence of Indian Point Unit 2 and other steam electric generating plants on the Hudson River Estuary, with emphasis on striped bass and other fish populations. Consolidated Edison Co. of New York, Inc.

McFadden, J.T. and Lawler, J.P. 1977. Supplement I to influence of Indian Point Unit 2 and other steam electric generating plants on the Hudson River Estuary with emphasis on striped bass and other fish populations. Consolidated Edison Co. of New York, Inc.

Pounder, J.R. and Rogers, T.D. 1980. The geometry of chaos: Dynamics of a nonlinear second-order difference equation. Bull. of Math. Bio. (In press)

Ricker, W.E. 1954. Stock and recruitment. J. Fish. Res. Board. Can. 11. 559–623

Silvert, W., and Smith, W.R. 1980. The responses of ecosystems to external perturbations. (In press)

Travis, C.D., Post, W.M., DeAngelis, D.L., and Perkowski, J. 1980. A compensatory
Leslie matrix model for competing species. (In press)

Walters, C.J. 1969. A generalized computer simulation model for fish population
studies. Trans. Am. Fish. Soc. 3: 505–512

Walters, C.J. and Ludwig, D. 1980. Effects of measurement errors on the assessment of stock-recruitment relationships. Can. J. Fish Oceans. (In press)

Wan, Y.-H. 1978. Computation of the stability condition for the Hopf bifurcation
of diffeomorphism on $\mathbb{R}^2$. SIAM J. Appl. Math. 34: 167–175

OPTIMAL MANAGEMENT OF OPTIMAL
FORAGERS

H. R. Pulliam
H. S. Colton Research Center
Museum of Northern Arizona
Route 4, Box 720
Flagstaff, Arizona   86001

Management models should reflect the reality that organisms will change
their behavior in response to environmental changes. Based on the evidence that the harvesting behavior of many animals approximates an
optimal behavior, it is shown here how this result may be used in a
model of predator-prey dynamics. Using this model it is shown that
equilibrium yields under human exploitation and qualitative stability
properties of predator-prey interactions depend critically upon the
behavior of the natural animal predators.

INTRODUCTION

Traditionally, most mathematical models of predator-prey interactions have
sacrificed realism and precision in order to achieve greater generality (Levins
1968). This approach has proven useful to evolutionary biologists interested in
community structure and evolution; however, such models have been of less use to
applied biologists interested in the management of particular natural resources.
Realistic modeling of predator-prey interactions is difficult because, to be
realistic, the behavior (= output) of a model must approximate the often very
complex behavior of the organisms being modeled. Both predator and prey organisms
change their behavior in complex ways in response to environmental changes, and the
ability to predict such changes in behavior may often be essential to sound management decisions.

Since humans want to be efficient harvesters of their natural resources, resource managers have become increasingly interested in applying optimization techniques to problems in resource management. The basic procedure has been to model
the dynamics of a natural system that includes man as a harvester of resources, and
then to use optimization techniques to find the harvesting regime that results in
the greatest yield. Animals other than man also need to be efficient harvesters of
their own food resources, and there is now some evidence that the harvesting behavior of many animals approximates optimal behavior in the sense of maximizing
yield to the individual animal harvester.

The current trend in human resource management is to aim for long-term (dynamic) optimization of resources. The maximization of long-term yield sometimes

requires prudence on the part of the individual harvesters because it is often necessary to sacrifice short-term gains in anticipation of future benefits. In the
absence of planning and cooperation, such prudence does not pay because prudent individuals lose resources to those who act to maximize their short-term gains. As
a result, in the natural world of non-human harvesters where the rules of resource
management are the product of individual Darwinian selection, prudent, long-term
optimizers are likely to lose the evolutionary race with short-term (static) optimizers. Accordingly, biologists have suggested that in nature animals forage so as
to maximize their short-term rate of resource (energy) intake.

As an example of the application of optimization techniques to the prediction
of animal behavior, consider the foraging behavior of a fish that eats zooplankton.
According to optimal foraging theory, if an individual fish behaved so as to maximize its short-term harvest of zooplankton, its relative preference for various
kinds of zooplankton would depend on their relative energy values. Prey value is
defined as the ratio of the caloric content of a prey item to its handling time.
Handling time is the time required to pursue, capture, and eat a prey item once
it has been located. A simple static optimization model shows that if a fish were
to maximize its rate of energy intake, it would eat the type of zooplankton with
the highest energy value every time this type is encountered. This model predicts
that the fish would only eat zooplankton of the type with the second highest energy
value if the abundance of the most valuable type dropped below a specified threshold. Likewise, the fish would only eat a type of zooplankton with yet a lower energy value if the combined abundance of the two most profitable types dropped below
another specified threshold. Qualitatively, then, the theory predicts that a fish
should eat a greater variety of zooplankton when the abundance of the more profitable types decreases, and conversely, it should specialize (eat fewer but more
profitable types) when the more profitable zooplankton are abundant.

Werner and Hall (1974) tested the quantitative predictions of optimal foraging
theory on bluegill sunfish by offering them three sizes of Daphnia as prey. They
found that when the large Daphnia were very abundant, the fish ate them exclusively,
but when the fish were offered fewer large Daphnia, they ate the medium size ones
as well. Similarly, when the large and medium size prey were rare, the fish ate
Daphnia of all three sizes, as predicted by the theory. Werner and Hall also found
that although the fish did not feed exactly as predicted by the theory, they came
very close to doing so. Similarly, in a study of Chipping Sparrows in southeastern
Arizona, I found that sparrows in the wild chose seeds so as to achieve about 95%
of the theoretical maximum rate of energy intake (Pulliam 1980). These results are
typical of a great many similar experiments on a variety of organisms (for reviews
see Pyke, Pulliam and Charnov 1977; Krebs and Davies 1979).

If, indeed, animals modify their behavior in response to changing prey abundance in a manner that is closely approximated by the predictions of optimization

models of behavior, then models of population interactions need somehow to account for this. Since animals are not perfect optimizers, models of population interactions based on models of optimal behavior, will only be approximate, but they will be closer to reality than traditional models that assume constant behavior in the face of changing environments.

ADDING BEHAVIOR TO POPULATION MODELS

A number of authors have realized the need to add realistic assumptions about behavior to models of population interactions. For example, Hassell and May (1974) and Sih (1979) have considered how the behavioral responses of prey to predator density affects the hunting success of predators. Oaten and Murdoch (1975) and Hassell and May (1974) showed how the functional responses of predators to the density of prey affected predator-prey dynamics. Oaten and Murdoch (1975) also considered the influence of predator switching behavior on the dynamics of predator and prey populations. In the analysis to follow, I consider how the changes in predator behavior predicted by optimal foraging theory can be incorporated into a model of predator and prey dynamics. I also attempt to show how the foraging behavior of natural animal predators might affect the harvesting decisions of human resource managers.

Consider the simple krill-whale model proposed by May, Beddington, Clark, Holt and Laws (1979). They model the dynamics of krill (X_1) and whale (X_2) populations with the equations

$$\frac{dX_1}{dt} = r_1 X_1 \left(1 - F_1 - X_1 - \nu X_2\right) \tag{1a}$$

and

$$\frac{dX_2}{dt} = r_2 X_2 \left(1 - F_2 - X_2/X_1\right) \quad , \tag{1b}$$

where F_1 and F_2 represent the human fishing rates on krill and whale, respectively. These authors did not present this as a realistic model of krill-whale interactions (J. R. Beddington and C. W. Clark, pers. comm.), but rather used the model to illustrate their central point that multi-species fisheries generally require different harvesting regimes than single species fisheries. Likewise, I do not pretend that this model gives an adequate description of krill-whale dynamics, but rather I use the model to make my point that optimal human management policies depend critically on the foraging behavior of animal predators.

In the equations of May et al., the parameter ν describes the rate at which krill are harvested by the predator, where krill abundance is measured in terms of the amount of krill required to support a single whale. Consequently, in the absence of human harvesting (i.e., $F_2 = 0$), the equilibrium whale population is given by $X_2 = X_1$. May et al. assume that the value of ν is always one, but in what follows, I assume that it can be either one or zero, depending on whether or not the whale is eating the krill. I also modify equation 1b to be

$$\frac{dX_2}{dt} = r_2 X_2 \left(1 - F_2 - \frac{X_2}{\nu X_1 + L}\right) \quad , \tag{2}$$

to allow for the possibility that the whale has a preferred prey that is eaten whether or not krill are being consumed. The abundance of the preferred prey, measured in units of the number of whales that can be supported, is given by L. An equation of dL/d_t could be added to the model to account for the changing abundance of the preferred prey, but for purposes of illustration, I assume that the preferred prey is of constant abundance and is always consumed by the whales. This would correspond to a natural situation where the abundance of the preferred prey is controlled at a constant level by some factor other than whale predation.

Because of the manner in which baleen whales forage, the energetic value of krill for whales depends on the abundance of krill. Thus, the decision of a whale to eat only the preferred prey or to eat both the preferred prey and krill should depend on the density of krill. Accordingly, I take the value of the parameter ν in equations 1a and 2 to be

$$\nu = \begin{cases} 0, & \text{if } X_1 \leq T \\ 1, & \text{if } X_1 > T \end{cases}$$

That is, the whale only eats krill if the abundance of krill is above some threshold T. The parameter T is the abundance of krill for which an expanded diet (i.e., one including krill) gives a greater rate of energy intake for a foraging whale than does a diet composed solely of the preferred prey.

Figure 1 shows how the threshold T changes the dynamics of the predator-prey interactions. In Figure 1a, I have assumed that there is no preferred prey ($L = 0$) so the whales always eat krill ($T = 0$). In Figure 1b, I have assumed that the preferred prey has abundance L and the whales only eat krill if X_1 exceeds the threshold T. The graphs are similar except that in Figure 1b, at low population levels of krill, where $X_1 < T$, the krill always increase in abundance and the whale populations have equilibrium abundance $X_2 = L$. From setting equations 1a and 2 both equal to zero, the equilibrium krill population size is

$$\hat{X}_1 = \frac{(1 - F_1) - \nu L (1 - F_2)}{1 - \nu (1 - F_2)} \quad . \tag{3}$$

If the value of $\hat{X}_1$ exceeds the threshold T, the equilibrium yield of krill to human harvesters is

$$\hat{Y}_1 = \frac{[(1 - F_1) - \nu L (1 - F_2)] \, r_1 K_1}{1 - \nu(1 - F_2)} \; . \tag{4a}$$

Similarly, the equilibrium yield of whales is

$$\hat{Y}_2 = \frac{[(1 - F+L) \, (1 - F_2)] \, r_2 K_2}{1 - \nu(1 - F_2)} \; . \tag{4b}$$

FIGURE 1a

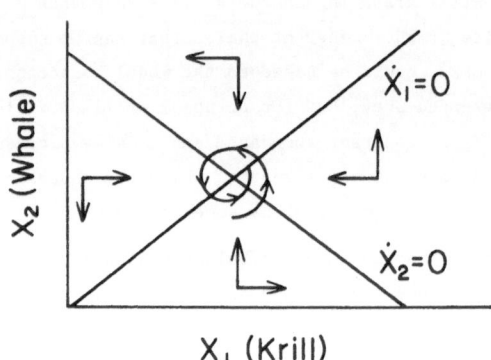

FIGURE 1b

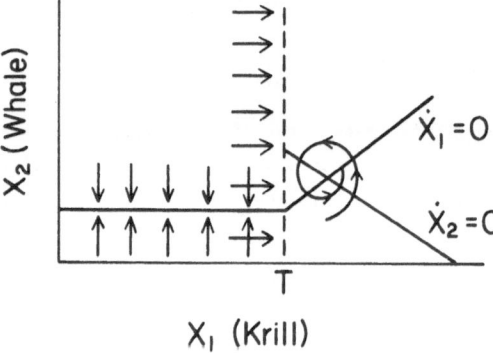

FIGURE 1. The switching behavior of an optimally foraging predator can be added to a simple model of predator-prey interactions (1a). In Figure 1b, the predator is assumed not to eat the prey unless the prey has abundance $X_1 > T$.

On the other hand, if the equilibrium abundance of krill ($\hat{X}_1$ as given in equation three) is less than the threshold T, then the value of ν is zero; that is, krill are not abundant enough to be included in the whales' diet. In this case, the equilibrium yield of krill and whales is simply

$$\hat{Y}_1 = (1 - F_1) \, r_1 K_1$$

and

$$\hat{Y}_2 = r_2 K_2 \ L \ (1 - F_2) \quad .$$

The point is that the foraging behavior of the whale, whether or not it eats krill at their equilibrium abundance, can make a very large difference in equilibrium yields, and therefore, should make a big difference in management decisions.

PREY CURVES WITH HUMPS

Some degree of realism can be added to a model of predator-prey interactions by adding a hump to the prey curve. Reasons for doing this are discussed in detail by Rosenzweig (1969). Briefly, the reasons are that at low prey population levels, there is often decreased productivity of the prey population and mating may be inefficient. At high prey population levels, there is food limitation, and even in the absence of predators, the prey population cannot grow at densities greater than the food imposed carrying capacity. The result is that the prey population can tolerate a greater amount of predation at intermediate prey density than at either extreme.

The qualitative stability properties of a predator-prey system with a hump in the prey curve are illustrated in Figure 2. If the angle between the predator curve and a tangent to the prey curve at the equilibrium is less than 90°, then the equilibrium point is locally stable. If this angle exceeds 90°, then the equilibrium point is locally unstable.

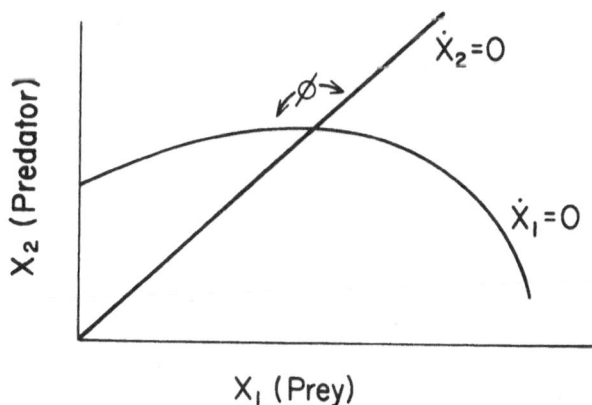

FIGURE 2. The local stability of the predator-prey equilibrium depends on the magnitude of the angle $\emptyset$.

Figure 3 shows two predator-prey systems differing only in the location of the threshold T, which is the density of prey at which the predators start to consume that kind of prey. In Figure 3a, the threshold is high and the equilibrium point is locally stable. In Figure 3b, the predator begins to include the prey in its diet at a lower prey abundance, and the equilibrium point is unstable, resulting in sustained predator and prey oscillations. Since the qualitative stability proper-ties of a predator-prey system may depend on the threshold T, a human management decision that affects T can directly affect the stability of the system.

FIGURE 3a

FIGURE 3b

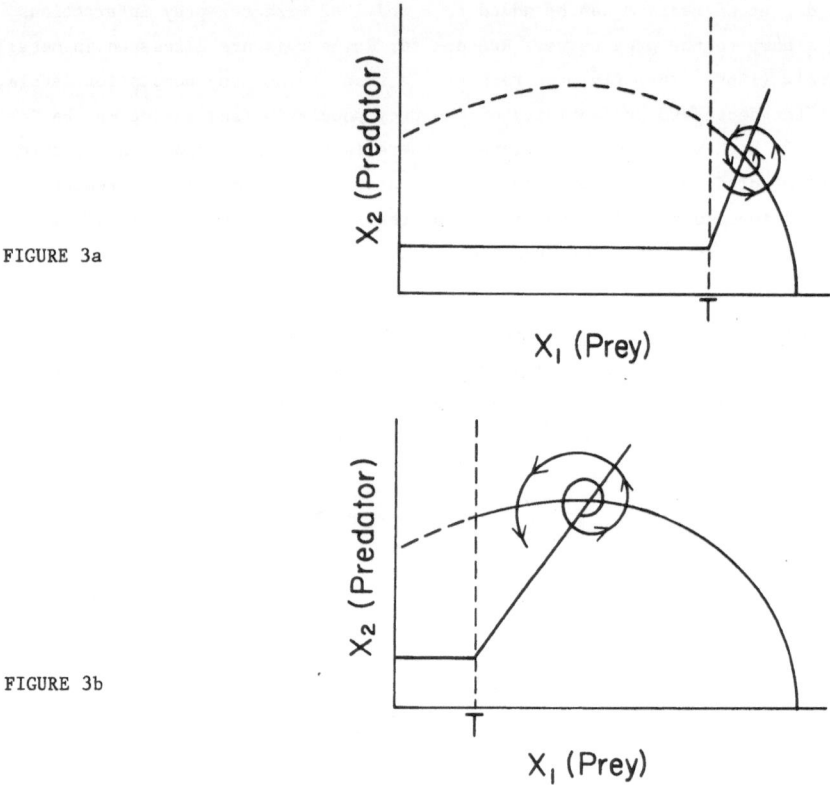

FIGURE 3. The qualitative stability properties of predator-prey interactions may depend on the threshold for inclusion of the prey in the predator's diet.

The conclusion to be drawn from this brief analysis is that both equilibrium yields and even qualitative stability properties of predator-prey interactions de-pend critically upon the foraging behavior of natural animal predators. Harvesting decisions, that are not based on a proper understanding of the flexible nature of predator behavior, may lead to a mismanagement of natural resources. Optimization models of foraging behavior may help resource managers make better decisions about human exploitation of natural predators and their prey.

REFERENCES

Hassell, M.B. and May, R. 1974. Ecology 43, 567-594

Krebs, J.R. and Davies, N.B. 1979. Behavioural Ecology: An Evolutionary Approach. Sinauer Press

Levins, R. 1968. Evolution in Changing Environments. Princeton Univ. Press

May, R., Beddington, J.R., Clark, C.W., Holt, S.J., and Laws, R.M. 1979. Science, 205, 267-277

Oaten, A. and Murdoch, W.W. 1975. Amer. Natur., 109, 289-298

Pulliam, H.R. 1980. Ardea, in press

Pyke, G.H., Pulliam, H.R., and Charnov, E.L. 1977. Quart. Rev. Biol., 52, 137-154

Rosenzweig, M.L. 1969. Amer. Natur., 103, 81-87

Sih, A. 1979. Jour. Anim. Ecol., 41, 79-84

Werner, E.E. and Hall, D.J. 1974. Ecology, 55, 1216-1232

MODELLING AND MANAGEMENT
OF FISH POPULATIONS WITH
HIGH AND LOW FECUNDITIES

B. S. Goh
Department of Mathematics
University of Western Australia
Nedlands, W.A., Australia

Usually a fish population with a high fecundity has a very variable recruitment. For this type of population, it is more realistic to abandon
the fiction that there is a well defined stock recruitment relationship.
Furthermore, the concept of maximum sustainable yield (MSY) is not applicable to this type of population. A good and realistic management
policy is to set the fishing mortality rate approximately equal to the
natural mortality rate and impose the optimal size limit for the specified fishing mortality rate. For a species with a low fecundity the
stock recruitment relationship is usually well defined. In this case,
it is possible to build a realistic model of the dynamics of the population, and the full power of stability theory and optimal control theory
may be used to formulate optimal management policies.

INTRODUCTION

The recruitment of many fish populations fluctuate greatly from year-to-year,
Cushing (1977), Hennemuth et al. (1979) and Doubleday (1980). In their study of
18 stocks of several species Hennemuth et al. (1979) found that the ratios of estimates for the smallest to the largest observed recruitments for various stocks
varied from 2.5:1 to 2700:1. Of these stocks, haddock has the most variable recruitment.

With such high variability the returns from using a realistic stochastic model
does not justify the cost in building, analysing and implementing such a model. A
similar situation occurs in the problem of driving a car through a busy city. It
is pointless to build a stochastic model incorporating the probabilities of meeting
other vehicles, pedestrians and constraints in order to programme the motion of a
car. It is simpler, more effective and cheaper to use feedback control and a model
of the motion of a car to determine the necessary course of action between any two
events. An event occurs when the car meets another vehicle or pedestrian or some
other impediment. Basically the approach is to formulate good deterministic policies for the period between any two unpredictable events. A similar approach
which avoids the unpredictable events was used successfully by Goh et al. (1975)
to manage tomato crops in greenhouses which are subjected to large variations in
total yield and price. In this paper, we shall examine a similar approach to

manage a fish population with large fluctuations in recruitment.

Living marine resources can roughly be divided into two classes. One class contains species with high fecundities, e.g., haddock, herring, plaice and prawns; and the other class contains species with low fecundities, e.g., whales, seals and penguins. This classification approximates the division of species into the r-strategists and the K-strategists, see Pianka (1974). In the classification of species into r and K strategists, the climate is supposed to play a central part in determining the type of strategy that is used by a species. But in the marine and terrestrial environments we can often find species with low and high fecundities living together in the same environment.

For our purpose, a species is said to have a high fecundity if an average mature female produces more than a hundred young animals or eggs per breeding season. If less than twenty young animals or eggs are produced each breeding season, it is said to have low fecundity. For many fish species, an average mature female can lay thousands of eggs per breeding season.

For a species with a high fecundity, it is difficult to have a precise relationship between a recruitment and the parent stock which produces the recruitment. The size of such a recruitment depends not only on the parent stock, but also on unpredictable environmental and biotic factors from the time of the birth of its members to the time of the recruitment. Typically the data for stock and recruitment for a species with high fecundity has a wide scatter. This is the case for a number of commercially important fish populations.

For a species with high fecundity and variable recruitment, I shall show that an incomplete modelling approach can explain adequately the main features of its catch effort data. This approach is biologically more realistic than that based on a stochastic logistic differential equation model. The latter has no time delay in the recruitment or any description of the age structure of the population. Therefore, the stochastic logistic model cannot be used directly and quantitatively to take advantage of an above average recruitment. Without a complete model, like the logistic model, it is <u>necessary</u> <u>to</u> <u>abandon</u> <u>the</u> <u>concept</u> <u>of</u> <u>maximum</u> <u>sustainable</u> <u>yield</u> <u>(MSY)</u>. In the incomplete modelling approach, the MSY policy is replaced by a set of rough guidelines whose object is to prevent the collapse of a fish stock, and within a set of guidelines, the optimal policy for maximum biomass yield is to impose an optimal size limit.

For a species with low fecundity, there is usually a well defined relationship between the recruitment and the parent stock when the population is sufficiently large. It is then possible to build a complete model, as in the case of a whale population. The mathematical tools of optimal control theory and stability theory are then available for the analysis of this type of model.

VARIABILITY IN STOCK RECRUITMENT AND CATCH EFFORT DATA

For a species with a high fecundity, the recruitment is usually very variable,
and it is not determined in a precise manner by the parent stock, see Figure 1.
In this case, it is better to abandon the fiction of a deterministic stock recruitment relationship. The task of predicting the recruitment from a given stock is
similar in kind to the problem of predicting the outcome of a lottery. For a stock
with highly unpredictable recruitment the pertinent management problem for maximum
biomass yield is to formulate the optimal harvesting policy for a given recruitment
and a given fishing mortality.

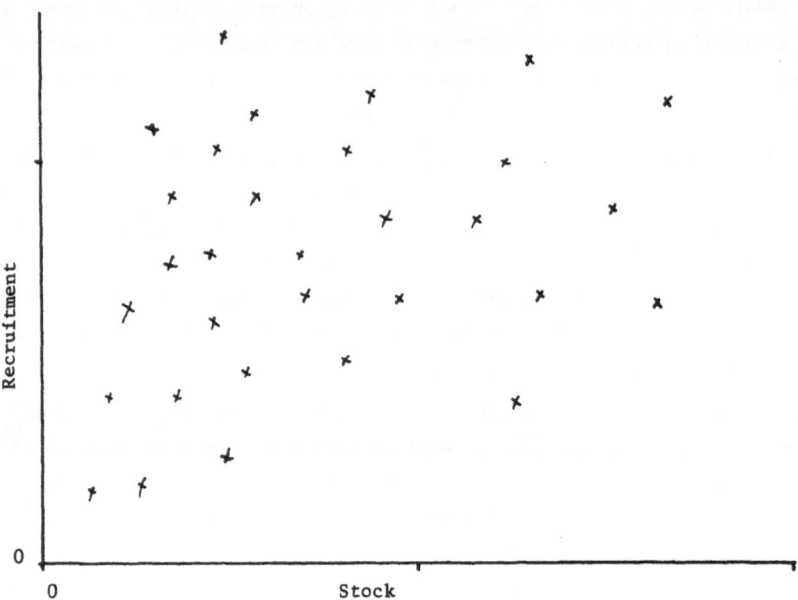

FIGURE 1. Typical stock recruitment data for a fish population with high fecundity

Figure 2 gives a typical set of data for the catch in tons per year versus the
fishing effort in hours for a fish population with highly variable recruitment.
There are two important qualitative features in this type of data. Firstly, the
data will fit a curve that is "parabolic" or asymptotic. Secondly, the scatter in
the data increases as the effort is increased. The fact that a "parabolic" type of
curve can be fitted to the data has led to the erroneous use of the logistic model
(the Schaefer model) in the management of fish populations with highly variable recruitment. We shall establish in this paper that the two main features of the catch

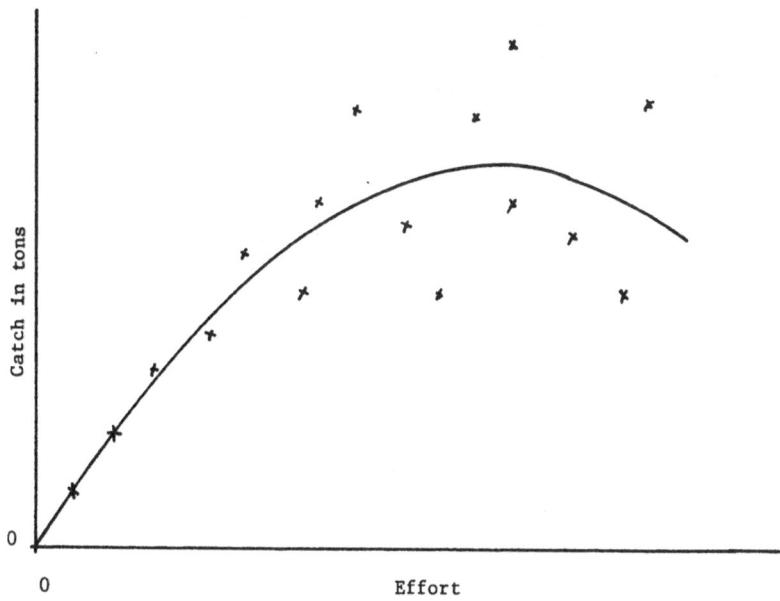

FIGURE 2. Typical catch effort data and fitted curve

versus effort data in Figure 2 can be explained using the incomplete modelling
approach.

A CATCH EFFORT MODEL

Under the tentative assumption that the recruitment is constant, we shall show
that the incomplete modelling approach will provide a catch effort model that has a
"parabolic" shape. We first show that in this approach, the relationship between
the number of fish that is caught and the effort is a function with a convex asymptotic shape. When we take into account the weight of an average fish at different
ages, then the catch in weight versus the effort is a function with a "parabolic"
shape.

For the incomplete modelling approach, we abandon the fiction that there is a
precise relationship between stock and recruitment. Following Beverton and Holt
(1957), we model a year class only from the age of recruitment t_r, which is the age
at which the year class enters an area where fishing is in progress. Without loss
of generality let t_r be also the age at which the fish is retained by the gear in
use. We assume knife-edged recruitment.

Let R denote the recruitment which is assumed to be a constant, M and F are the instantaneous natural and fishing mortalities and T is the life span of an average fish. If N(t) is the number of fish in the year class at time t, then

$$\dot{N} = -(M + F)N, \qquad N(t_r) = R \quad . \tag{1}$$

Under the assumption that F is a constant the number of fish in the year class at time t is

$$N(t) = R \exp -(M + F)(t - t_r) \quad . \tag{2}$$

With constant recruitment, the number of fish that is caught each year from a population which contains the recruited fish of all ages, is equal to the number of fish that is caught from a single year class throughout its life span. Thus the number of fish caught each year in the steady state condition is

$$N_c = \int_{t_r}^{T} F\, N(t)\, dt$$
$$\approx FR/(M + F) \quad , \tag{3}$$

because N(T) is negligible. Under the assumption that F is proportional to the fishing effort, the relationship between N_c and effort is a function which is convex and asymptotic, see Figure 3.

In an exploited population the proportion of old fish in a population decreases as the fishing effort is increased. Under heavy exploitation the population may, for practical purposes, contain only a single year class that is recruited in the same year. Since old fish are heavier than young fish, we can expect that beyond a certain point the total catch in weight decreases as effort is increased. This is in fact the case.

Under constant recruitment and in the steady state situation the total catch in weight each year from a population containing multiple age classes is

$$Y = \int_{t_r}^{T} FW(t)N(t)\, dt \quad . \tag{4}$$

$$\Rightarrow \frac{dY}{dF} = \int_{t_r}^{T} [1 - F(t - t_r)]W(t)N(t)\, dt \quad . \tag{5}$$

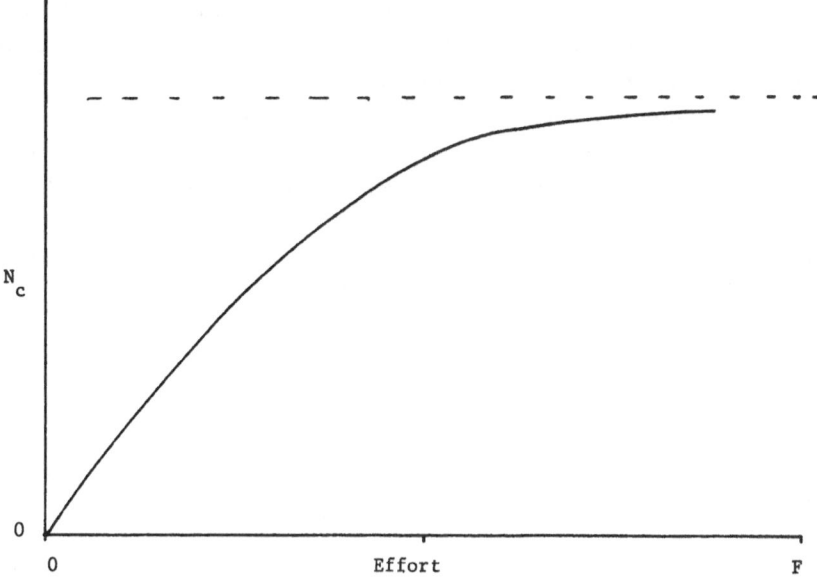

FIGURE 3. Qualitative behaviour of number of fish caught versus effort

From (5) it is clear that dY/dF is negative if F is sufficiently large because $N(t) > 0$ and $W(t) > 0$. Under the assumption that the fishing mortality F is proportional to effort Figure 4 provides an example on how the catch in weight varies as effort is increased.

We make the important conclusion that a linear population model in the complete modelling approach can produce a "parabolic" type of relationship between catch and effort. It is unnecessary and biologically unrealistic to use a logistic differential equation model (the Schaefer model) to establish a parabolic relationship between catch and effort.

In the real world the recruitment of fish populations with high fecundity varies from year to year. For some fish populations the recruitment may vary by a thousand fold. We shall now consider how the variations in recruitment affect the catch versus effort data.

When the effort for harvesting is small, a population containing many year classes will damp out the year-to-year variations in the recruitment. The recruitments are like varying inputs of water into a reservoir. The catch in a particular year will depend on the effort and the total population in that year. If the total population contains many year classes and the total population, in the absence of harvesting, is relatively constant in spite of the varying recruitment, then for low harvesting effort the catch effort data would follow closely a catch effort model which is developed under the assumption of a constant recruitment. With high fishing effort the total population would contain only one or two significant year classes. In this case, the total catch in a particular year will depend directly on the current recruitment and/or on an exceptional recruitment. Thus the total catch will vary from year to year even if the effort is maintained at the same level. This is because the catch tracks the varying recruitment.

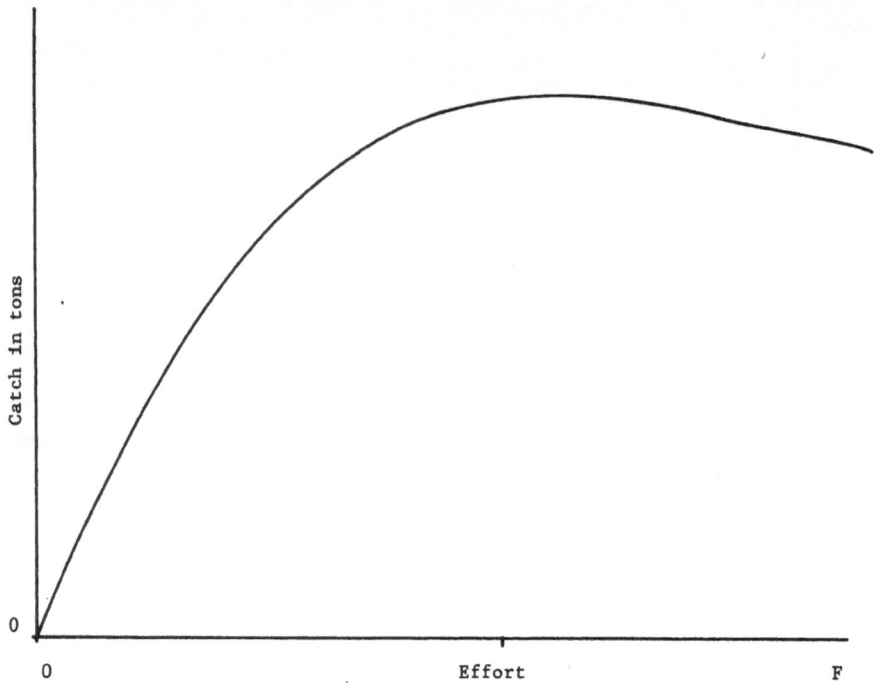

FIGURE 4. Catch versus effort for model of North Sea plaice with $t_r=7.3$ and $M=0.1$

We conclude that a linear model in the incomplete modelling approach can explain satisfactorily and realistically the two main qualitative features of a typical catch effort data for a fish population. This conclusion has important consequences for the management of such a fish population.

A GOOD HARVESTING POLICY

At this point, let us accept the proposition that the incomplete modelling approach provides a realistic approach to study the dynamics of a fish population with a high fecundity. An immediate and important consequence of this proposition is that we must abandon the concept of maximum sustainable yield (MSY) for this type of fish population. In the incomplete modelling approach the question on whether or not there is a relationship between stock and recruitment is circumvented. Thus there is no description of the long term dynamics of the fish population.

After throwing out the MSY concept we must ask what are the pertinent management questions? Firstly, we can ask what is the optimal harvesting policy for maximum biomass yield if the fishing mortality rate F is fixed? If F is fixed, the optimal policy is to prescribe an optimum size limit. This requires that only fish which are larger than the optimum size limit, Goh (1977, 1980) should be harvested. This policy is optimal even when the recruitment is highly variable. Using the policy of imposing an optimum size limit the optimum yield can be increased indefinitely by increasing the value of F. But in practice, F cannot be increased indefinitely without the danger of destroying the recruitment altogether. Thus the absolutely best harvesting policy for biomass yield does not exist. We can only formulate a reasonably good policy which provides a good biomass yield without taking a serious risk in the destruction of the fish population.

Therefore, we have to specify the value of F in some other way. Taking into account some economic factors, pulse fishing may be desirable, in which case F is a function of time. But let us adopt the view that for steady employment it is desirable to fix F at a constant value. Until better and more detailed knowledge of the dynamics of a fish population with high fecundity is available, I would like to suggest that it should be managed using the following scheme for classifying the management status of such a population. If $0 \leq F \leq 0.5$ M, then the population is said to be underexploited. If 0.5 M $\leq F \leq 2$ M, then the population is said to be exploited at a desirable level. If 2 M $\leq F \leq 4$ M, the population is said to be overexploited. If $F \geq 4$ M, the population is said to be highly overexploited. Given F, the corresponding optimum size limit should be imposed.

This classification may be refined by taking into account the number of significant year classes that is present in the population if it is not exploited. If the latter has only one or two year classes, it is probably wise to choose F = M . If it has many year classes, it would be safe and worthwhile to set F = 2 M . Furthermore, F should be temporarily increased to take advantage of an exceptional recruitment to the population.

The rationale behind this management classification is as follows: Firstly, a species with a high fecundity has evolved to sustain high natural mortality rates before recruitment and a certain mortality rate after recruitment. Therefore, the recruited part of the population would probably be able to sustain an increase in the instantaneous mortality rate from M to 3 M . If the fishing mortality rate F is too large, there may be only one or two significant year classes present in the harvested population. This could cause the destruction of the population if the recruitment should fail in a number of consecutive years, In other words, an over-exploited population is not robust to such failures in recruitment.

HARVESTING SPECIES WITH A LOW FECUNDITY

For a species with a low fecundity we would expect a fairly well defined re-lationship between stock and recruitment. This is a K-strategist and typically a parent will devote a large amount of resource to ensure the survival of a single offspring to the adult stage. However, density dependent mortalities of the off-springs will normally occur at high densities, because of competition for food and space. With a well defined stock recruitment relationship, it is then possible to build a deterministic model of the complete dynamics of a population if the stock is relatively large.

We can distinguish at least two types of species with low fecundities, those with a complex social structure like the sperm whale and those with a relatively fixed sex ratio like the baleen whales. To manage a population with a complex social structure, it is necessary to build a model incorporating both the male and female subpopulations and a description of the social behaviour like the keeping of a harem in the case of the sperm whale.

For a species like a baleen whale, a relatively simple model describing only the female subpopulation may be adequate to describe the dynamics of the population. The full power of stability theory, see Goh (1980), and optimal control theory, see Clark (1976), may then be used to analyse such a model.

The baleen whale model is of the form

$$N(t + 1) = sN(t) + R[N(t - k)] , \qquad (6)$$

where $N(t)$ is the number of· mature and reproductive females at time t, s the surviv-al rate of a mature female and $R[N(t - k)]$ is a stock recruitment function. It de-scribes the number of female recruits to the population at time $(t + 1)$ which was due to the parent stock at time $(t - k)$. For the Antarctic fin whale, the unex-ploited population model is globally stable. The concept of maximum sustainable yield is applicable to model (6).

REFERENCES

Beverton, R.J.H. and Holt, S.J. 1957. On the Dynamics of Exploited Fish Populations. Fish. Invest. London Ser. 2, Vol. 19, 1-533

Clark, C.W. 1976. Delayed-recruitment Model of Population Dynamics with an Application to Baleen Whale Populations. J. Math. Biol., 3, 381-392

Cushing, D.H. 1977. The Problems of Stock and Recruitment. In, Fish Population Dynamics, Gulland J.A., ed. Wiley, New York, 116-133

Doubleday, W.G. 1980. Coping with Variability in Fisheries. FAO Fisheries Report No. 236, 131-139

Goh, B.S. 1977. Optimum Size Limit for a Fishery with a Limited Fishing Season. Ecological Modelling, 3, 3-15

Goh, B.S. 1980. Management and Analysis of Biological Populations. Elsevier Scientific Publ. Co., Amsterdam

Goh, B.S., Peng, W.Y., Vincent, T.L. and Riley, J.J. 1975. Optimal Management of Green House Crops. Hortscience 10(1), 7-11

Hennemuth, R.C., Palmer, J.E. and Brown, B.E. 1979. Recruitment Distributions and their Modality. Part 1. Description of Recruitment in 18 selected Fish Stocks. ICES C.M. (mimeo) 1-57

Pianka, E.R. 1974. Evolutionary Ecology. Harper and Row Publ., New York.

ADAPTIVE IDENTIFICATION OF MODELS
STABILIZING UNDER UNCERTAINTY

Janislaw M. Skowronski
Department of Mathematics
University of Queensland
St. Lucia, Brisbane, Qld 4067
Australia

The harvested ecosystem is seen as a "black-box" with the input-output data measured within a given noiseband. An attempt is made to <u>explain</u> the adaptive state or parameter identification method widely used in the Linear Control Theory and to <u>introduce adjustments</u> allowing application of this method to ecological systems which are nonlinear. Finally a fairly general-type of nonlinear dynamic model with adaptively identified variables is conjectured for managing the system towards a stable growth within desired regions of the state space. In all above sufficient conditions for the identification and stabilizing are obtained via the Liapunov Direct Method.

INTRODUCTION

The analytic (non-control) models in ecodynamics lack in providing options for man-interference in the process of nature which then may be either too slow or not at all directed wherever we wish it to go. Moreover such models, if used for management, give dangerously misleading picture. Brauer-Soudack-Jarosch (1976) argued the latter point successfully and lectures as well as discussions during this Workshop (cf. Beddington-Botkin-Levin, in this Volume) reinforced this view. It is argued that management models must be far more realistic ecologically. This requires more complex models (Botsford, in this Volume) and a greater variety of them (Beddington, in discussion). The above in turn requires more data which quite often are not available for various reasons - the cost being not the least important. However, there is no alternative but to keep on modeling. In fact, Clark (in this Volume) says that models are especially useful where there is not enough information. Then however we should make them flexible enough to search on-line for the system identity (state variables and parameters) time-instantaneously matching the available data. This is known in control theory as adaptive identification with the model called an identifier. The difficulty lies in that this method, although applied very widely, so far has been developed mainly for linear systems while the realistic approach to ecodynamics must be nonlinear and nonlinearizable (several equilibria, ... etc). Hence the method calls for extension. In all of this we must be prepared to handle also a lot of noise due to the uncertainties in modeling and the inaccuracy in data measurement. Further, the identifier must be made capable of securing some objectives for the manager, e.g., stabilize the growth of the populations in a prescribed region of the state space or avoid another region (non-vulnerability), obtain some type of optimal behaviour, ... etc.

There are at least two ways of constructing the identifier we want. We may make a giant model to include almost everything needed and then try to prove its consistency with the data as well as its stability, ... etc., – an enormous task if not impossible. The practical alternative is to introduce a set of sufficient conditions for the adaptive identification and for the desired objectives, together with an identifier which is general enough to be applicable. Then we may take one by one from the variety of particular-property orientated hypothetical models (candidates) reasonably consistent with ecological reality or a combination of such models, all covered by the said general identifier, and check it against these sufficient conditions. This produces successive augmentation of the model with some characteristics and parameters known, some still to be identified at each stage.

What one means by "general" identifier depends obviously very much on the situation. Since our aim in this paper is to open a discussion on the method rather than to produce specific results - simple examples followed by an almost arbitrary format (except for obvious mathematical assumptions) of the right hand sides of the dynamic equations, cf. (4.1), seems to be the most suitable background for the study.

In Section 2 we explain the elementary concepts of the method on its presently popular linear stage. In Section 3 a simplified version (noiseless, autonomous) of our nonlinear identifier is introduced with easy to find sufficient conditions. Then we use the latter to confirm a particular candidate: the logistic model. Although much criticised it is still the most familiar model and thus a very good tool for illustrating the method. Section 4 conjectures a general noise-robust identifier together with relevant sufficient conditions. The applicability of it is still to be verified.

ELEMENTARY CONCEPTS

Let us assume our "black-box" linear:

$$\dot{x}_p = A_p x_p + B_p u + R(t) , \qquad (2.1)$$

$$y = C x_p , \qquad (2.2)$$

with the vector of state variables $x_p \in R^N$, the control vector $u \in R^r$, $r \leq N$, the output vector $y \in R^n$, $n \leq N$ and the time $t \in J_o = [t_o, \infty)$ where $t_o \in R$ is an initial instant. Further, A_p, B_p, C are constant matrices of order $N \times N$, $N \times r$, $n \times N$

respectively representing parameters and R is the N × 1 input matrix. Usually the values of u would be generated by some control program which is expected to implement some desired objectives (stabilization, optimal behaviour, ..., etc.), but in this Section, we wish to concentrate on identification only, hence let us assume the objectives satisfied and u given. Also, since we do not want to concentrate on the read-out problem (2.2) we may as well assume C given.

In this section the blackness of our box is reflected in the fact that R and y have been measured while either x_p, or A_p, B_p are <u>unknown</u> except for that the values of x_p and the values of the constant parameters of A_p, B_p must be located in the <u>known</u> bounded sets Δ_o, A, B, respectively. Now let us form the system

$$\dot{x}_m = (A^o_m - KC)x_m + Ky + B^o_m u + R(t) \tag{2.3}$$

where either $x^o_m = x_m(t_o)$, or A^o_m, B^o_m are arbitrary guesses from Δ_o, A, B respectively and K is some constant N × n matrix to be designed for the sake of the state-identification part of the problem. Note that Ky is now a part of control with the program (2.2), while R(t) is the same matrix as in (2.1).

Since the sets Δ_o, A, B are bounded and known, the value of x^o_m or the values of the parameters of A^o_m, B^o_m may not be further from these of $x^o_p = x_p(t_o)$, A_p, B_p than within the known diameters of Δ_o, A, B, respectively. We shall postulate now that in each case the distances will decrease to zero asymptotically as $t \to \infty$. Let us make $A_m(t)$, $B_m(t)$ time dependent with $A^o_m = A_m(t_o)$, $B^o_m = B_m(t_o)$. Further let us substitute (2.2) into (2.3) and subtract (2.1). We have

$$\frac{d}{dt}(x_m - x_p) = [A_m(t) - KC](x_m - x_p) + [A_m(t) - A_p]x_p + [B_m(t) - B_p]u \quad .$$

Introduce now the <u>state error</u> $e(t) = x_m(t) - x_p(t)$, the <u>parameter misalignment</u> matrices $A(t) = A_m(t) - A_p$, $B(t) = B_m(t) - B_p$ and denote $K = A_m(t) - KC$. It produces the so called <u>error-equation</u>:

$$\dot{e} = Ke + Ax_p + Bu \tag{2.4}$$

which is fundamental for the linear identification.

According to our postulate, for the identifier (2.3) to work, it is required that either $e^o = e(t_o)$, or $A^o = A(t_o)$ and $B^o = B(t_o)$ successively from Δ_o, A, and B, imply correspondingly

$$e(t) \to 0, \text{ as } t \to \infty \tag{2.5}$$

$$A(t) \to 0, \; B(t) \to 0, \text{ as } t \to \infty \; . \tag{2.6}$$

Then obviously for sufficiently large t, no matter where the values of x_p and of the parameters in A_p, B_p have been, the differences will be small enough to secure the identity between the system and the model.

Let us first look at the simpler case of state identification when the parameters have been identified already: $A_m(t) \equiv A_p$, $B_m(t) \equiv B_p$. Then (2.3) becomes the known <u>state</u> (only) <u>Luenberger observer</u>:

$$\dot{x}_m = (A_p - KC) \, x_m + Ky + B_p u, \tag{2.3}'$$

with the error equation

$$\dot{e} = (A_p - KC)e \; . \tag{2.4}'$$

It then suffices to select K such as to secure $(A_p - KC)$ having negative real parts thus yielding the asymptotic (exponential) stability of the equilibrium: $e(t) \to 0$ at a certain exponential rate as $t \to \infty$. The "more negative" these real parts are, the faster this rate becomes. To obtain a derived accuracy of the approximation of our black-box it remains to propose a suitable K.

When the parameter identification must take place: $A_m(t) \neq A_p$, $B_m(t) \neq B_p$, but $x_p(t)$ is known the Liapunov formalism is employed to produce (2.6), i.e., the asymptotic stability of the equilibrium in the space $E = \{e,A,B\}$. Following Kudva-Narendra (1972) it is possible to obtain a Liapunov function which is a positive definite quadratic form in E:

$$V(e,A,B) = \frac{1}{2} \; (e^T Pe + \sum_{i=1}^{N} A_i^T A_i + \sum_{i=1}^{r} B_i^T B_i) \tag{2.7}$$

where $P = P^T$ is positive definite matrix and A_i, B_i are the i-th columns of A,B respectively. Then

$$\dot{V} = -e^T Qe + e^T P\mu + \sum_{1}^{N} (\dot{A}_i^T A_i + \dot{B}_i^T B_i) \tag{2.8}$$

where $-Q = \frac{1}{2} \; (K^T P + PK)$, $Q = Q^T$, $\mu = Ax_p + Bu$ and Q is any positive definite symmetric matrix.

Substitution into (2.8) shows that if

$$\dot{A}_i^T = -e^T P x_{pi} \ , \qquad i = 1, \ldots, N$$

$$ \tag{2.9}$$

$$\dot{B}_i^T = -e^T P u_i \ , \qquad i = 1, \ldots, r$$

then

$$\dot{V} = -e^T Q e \tag{2.10}$$

which is negative semi-definite and produces stability. The required asymptotic stability is obtained by "intensifying" the control u in a specific way which requires a rather long exposition, cf. Narendra-Kudva (1974), and is beside our point here. The equations (2.9) are known as the adaptation law and represent the so called <u>adaptation mechanism</u> of the model. The block scheme is shown in Figure 1.

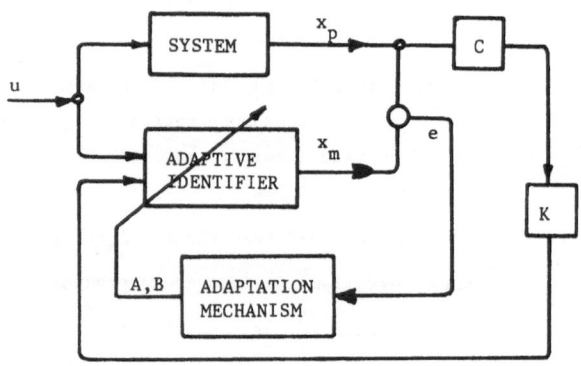

FIGURE 1. Block scheme for linear identification

SIMPLIFIED NONLINEAR CASE

We begin our nonlinear study with a simplified case of an autonomous noiseless harvested eco-(black)-box:

$$\dot{x}_p = f_p (x_p, \lambda_p) - h \ , \tag{3.1}$$

$$y = g(x_p) \ , \tag{3.2}$$

where $x_p \varepsilon R_+^N$ (the positive quadrant of R^N), $h \varepsilon R_+^N$ is the harvest-control vector, and $\lambda_p \varepsilon \Lambda \subset R^\ell$ is the constant tuple of parameters — for instance if (3.1) is logistic, λ_{p1} may be the intrinsic growth rate, λ_{p2} the carrying capacity, ..., etc. Again in general, $h(t)$ is to be implemented by a program (adaptive feedback controller) designed to produce specified objectives. In this secion, for the same reason as before, we assume all the objectives (except identification) already attained and thus h known. One of the assumed objectives would have to be the Lagrange stability of (3.1) in a preassigned $\Delta_o \subset R_+^N$ ($\equiv$ invariance of Δ_o under the trajectores of (3.1)) producing $x_p(t) \varepsilon \Delta_o$, $t \geq t_o$.

Presently, <u>we know</u> h, y, $f_p(\cdot)$, $g(\cdot)$ and <u>we do not know</u> x_p, λ_p, but Δ_o, Λ are given. The adaptive identifier is now proposed in the form

$$\dot{x}_m = f_m(x_m, y, \lambda, h) , \qquad (3.3)$$

with $f_m(\cdot)$ to be designed, $x_m(t) \varepsilon \Delta_o$, $t \varepsilon J_o$, and the misalignment vector $\lambda(t) = \lambda_m(t) - \lambda_p$ produced by an <u>adaptation mechanism</u> which uses both y and x_m. The latter is identified with the output of (3.3) since the model is observable by definition. The mechanism may be represented by a program generating a suitable function $\lambda(\cdot)$. In particular, but not necessarily, it can be the dynamics corresponding to (2.9) defined to suit some Liapunov condition — as in Section 2:

$$\dot{\lambda} = f_a(y, x_m), \quad \lambda(t_o) = \lambda^o \varepsilon \Lambda_o \subset \Lambda . \qquad (3.4)$$

Whatever the case, λ must be bounded in Λ. Note that $\dot{\lambda} \equiv \dot{\lambda}_m$ and (3.4) serves also as the law of motion for $\lambda_m(t)$.

The system $\{(3.1), (3.2), (3.3)\}$ replaces now the error equation (2.4). The functions $f_p(\cdot)$, $g(\cdot)$ and $f_m(\cdot)$ must be such as to produce unique solutions to this system from $\bar{\Delta}_o \times \bar{\Delta}_o = \bar{\Delta}_o^2 \subset R_+^{2N}$. Let $x(t) = (x_p(t), x_m(t))$, $t \varepsilon J_o$ with $x^o = x(t_o)$ and let the solution $x(x^o, \cdot) : J_o \to R_+^{2N}$ through $x^o \varepsilon \bar{\Delta}_o^2$ be represented by the trajectory $x(x^o, J_o) = \{x(x^o, t) | t \varepsilon J_o\}$ in R_+^{2N}.

The identification (in this Section) is considered completed, if $f_m(\cdot)$ is found such that the equilibria of (3.3) and (3.1) coincide, and given y, x_m, h there is $f_a(\cdot)$ of (3.4) and the corresponding λ such that Liapunov conditions may be satisfied to secure the following: $x^o \in \Delta_o^2$ and $\lambda^o \in \Lambda_o$ imply

$$x(x^o,t) \to M, \text{ as } t \to \infty , \qquad (2.5)'$$

$$\lambda(t) \to 0 \text{ as } t \to \infty , \qquad (2.6)'$$

where $M = \bar{M} \cap \bar{\Delta}_o^2$ and $M = \{x \in R_+^{2N} | x_p = x_m\}$ is the diagonal set of R_+^{2N}. Then no matter what x_p, λ_p have been before, for sufficiently large t the identity between (3.1) and (3.3) will be secured. The problem is by no means unique. There will be a large class of $f_m(\cdot)$, $f_a(\cdot)$ yielding the identification even when the corresponding Liapunov function is established. Granted the invariance of

$\Delta_o^3 = \Delta_o^2 \times \Lambda \subset R_+^{2N} \times R^\ell$ which follows from the assumed Lagrange stability and (3.4) and thus also granted the invariance of $\bar{\Delta}_o^3$ (note that Δ_o^2 and Λ are bounded) we need, as known cf. Yoshizawa (1966), a smooth function V: $\bar{\Delta}_o^3 \to R$ such that

(i) $V(x,\lambda) = 0$, $x \in M$, $\lambda = 0$

(ii) $V(x,\lambda) > 0$, $x \notin M$, $\lambda \neq 0$ (3.5)

(iii) $\nabla V(x,\lambda) \cdot f(x,\lambda,h) < 0$, $x \notin M$, $\lambda \neq 0$,

where $f(x,\lambda,h) = (f_p(x_p) - h, f_m(x_m,\lambda,h), f_a(y, x_m))$. From the known geometric interpretation of (iii) we conclude that the further below zero is the derivative, the faster $x \to M$, $\lambda \to 0$ as $t \to \infty$.

To compare the linear method of the Section 2 with the present amendment, let us see how the latter works for instance for the Luenberger observer. We let (3.1), (3.2) with λ_p known be specified by (2.1), (2.2) with A_p, B_p known and let the candidates for $f_m(\cdot)$, $\lambda = 0$ and $V(\cdot)$ be defined by (2.3)' and $V(x) = (x_m-x_p)^T (x_m-x_p)$ respectively. The condition (iii) becomes

$$\dot{V} = (\dot{x}_m - \dot{x}_p)^T (x_m - x_p) + (x_m - x_p)^T (\dot{x}_m - \dot{x}_p) .$$

Substituting (2.4)' we obtain

$$\dot{V} = -(x_m - x_p)^T Q(x_m - x_p)$$

where $-Q = (A_p - KC)^T + (A_p - KC)$. With the Luenberger K the matrix Q is positive definite. This makes the conditions (3.5) satisfied and the candidates confirmed.

Moreover we observe that $V(x(t)) = V(t)$ decreases at similar exponential rate as the solutions of (2.4)'. Hence the match between Luenberger and the amendment is not only qualitative but also quantitative.

It seems instructive to illustrate this Section on the simple specific case of the scalar logistic equation. We shall consider

$$\dot{x}_p = x_p - \lambda_p x_p^2 - h \quad . \tag{3.1}'$$

The other parameter with x_p may be reduced to 1 if λ_p is taken as a ratio. A vector equation, say only two dimensional, would lead immediately to the system (3.1), (3.3) being four dimensional and thus the example obscuring the method rather than explaining it. However there is a penalty for single dimension of (3.1)'. The output vector y cannot be made less dimensional than x_p and thus the state-identification procedure becomes trivial. We may as well assume the state to be known and take the read-out function $g(\cdot)$ defined by

$$y = 2x_p \quad . \tag{3.2}'$$

Then we let

$$\dot{x}_m = -x_m - \lambda_m x_m^2 + y - h \quad , \tag{3.3}'$$

$$\dot{\lambda} = (x_m - \tfrac{y}{2})(\tfrac{y}{2})^2 \quad ; \tag{3.4}'$$

and $V = \frac{1}{2}(x_m - x_p)^2 + \frac{1}{2}\lambda^2$, whence $\dot{V} = (x_m - x_p)(\dot{x}_m - \dot{x}_p) + \lambda\dot{\lambda}$.

We calculate $\dot{x}_m - \dot{x}_p = -x_m - \lambda_m x_m^2 + 2x_p - x_p + \lambda_p x_p^2$

$= -(x_m - x_p) - (\lambda_m x_m^2 - \lambda_p x_p^2) = -(x_m - x_p) - \lambda_m(x_m^2 - x_p^2) - x_p^2(\lambda_m - \lambda_p)$,

thus obtaining

$$\dot{V} = -(x_m - x_p)^2 - \lambda_m(x_m - x_p)^2(x_m + x_p) - x_p^2(x_m - x_p)(\lambda_m - \lambda_p) + (\lambda_m - \lambda_p)\dot{\lambda} \quad .$$

Since $x_m + x_p > 0$, for suitable λ_m^o selecting $2x_m^o > y(t_o)$ by (3.4)' the Liapunov derivative is negative definite. In view of the above, the conditions (3.5) are satisfied thus yielding the desired identification.

Unfortunately, both the linear adaptive identifier and our nonlinear generalization so far introduced share basic deficiency namely the asymptotic approach

$(x_m - x_p) \to 0$, $\lambda \to 0$, as $t \to \infty$ implied by either of them. With that, the physical problem is not exactly well-posed. In the first place, we will never have $x_m = x_p$, $\lambda = 0$. But even assuming that one agrees with some approximation there is no such thing as the best approximation since one can always improve by choosing better. Obviously, the same argument applies to possible minimizing of the time interval for reaching M. Finally the defect which is essential for nonlinear system – this identifier being asymptotic does not differentiate between equilibria – hence it can be used only locally, in a neighborhood of some equilibrium. In the next section we shall attempt to avoid these shortcomings.

THE GENERAL CASE

In this section we do not assume any objectives attained. In particular, stabilization will be part of the investigated problem. Hence there will be a program required for h which is feedback related to the identified x_m and adaptively related to the identified misalignment λ. The block-scheme of the problem is shown in Figure 2. It encloses two major blocks: the eco-box and the identifier, as well as two managing agents: the adaptation mechanism and the adaptive control program. Our present eco-box is the nonautomomous version of (3.1):

$$\dot{x}_p = f_p(x_p, \lambda_p, t) - h , \qquad (4.1)$$

with the t-explicite under $f_p(\cdot)$ representing the feedforward input, see Figure 2. Note that λ_p = const.

We admit noise in y due to uncertainty in measurements. In one way, the noiseband may be obtained through a set-valued read-out function $G : R_+^N \times R \to$ family of Y(t), $t \in J_o$, defined by $y(t) \in G(x_p(t), t) = Y(t)$, where Y(t) are compact subsets of $Y \subset R_+^N$. This produces

$$y \in G(x_p, t) . \qquad (4.2)$$

Alternatively we may consider directly the t-family of the sets Y(t). In any case $\{Y(t) | t \in J_o\}$ is given.

The output noise pollutes the identifier (see Fig. 2) directly, and also indirectly through the adaptation mechanism and consequently the control program. The non-autonomous versions of (3.3) and (3.4) are subsequently

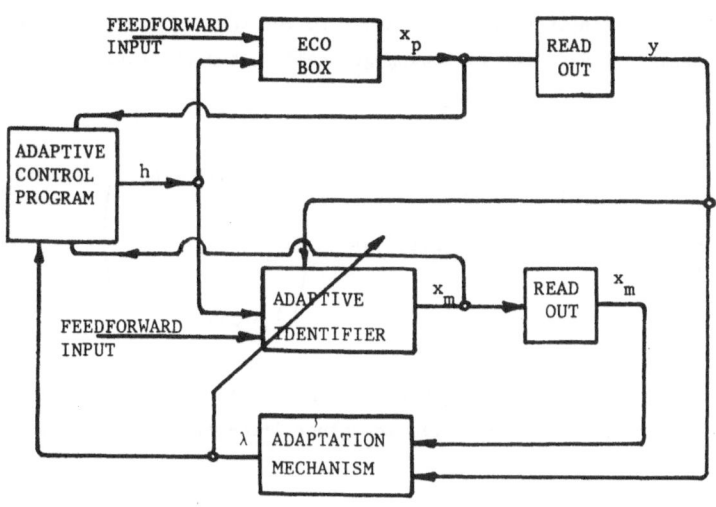

FIGURE 2. Block scheme for the general case

$$\dot{x}_m = f_m(x_m, y, \lambda, h, t) \ , \tag{4.3}$$

$$\dot{\lambda} = f_a (x_m, y, t) \ , \tag{4.4}$$

and must be now <u>considered for all</u> $y \in Y(t)$. Then (4.4) generates sets
$\Lambda(t) \subset \Lambda$ of $\lambda(t)$. This, apart from influencing solutions of (4.3), makes up for a
set valued control program. We let $P : R_+^{2N} \times R^\ell \to$ family of $P(t)$, $t \in J_o$, in given
compact $P \subset R_+^N$, defined by

$h(t) \in P(x_p(t), x_m(t)) = \{p(x_p(t), x_m(t), \lambda(t)) | \lambda(t) \in \Lambda(t)\} = P(t)$.

Except for the inclusion of objectives (selection of h) the philosophy of our
study is similar as previously: we know $f_p(\cdot)$, Y, P, we do not know x_p, λ_p within
given (desired) Δ_o, Λ_o respectively and we design (4.3), (4.4) and $P(\cdot)$, implying
$x_m(t)$, $\lambda(t)$ which produce the identification (up to the specified accuracy) and the
objectives.

The system $\{(4.1), (4.3), (4.4)\}$ subject to (4.2) and all $h \in P$ suits the
discussion for most practical purposes. Thus, it will also be used in our

sufficient conditions. However, in order to get there formally by the shortest
path we shall introduce the following generalized representation. Let us form the
vectors:

$$
z(t) = \begin{pmatrix} x_p(t) \\ x_m(t) \\ \lambda(t) \end{pmatrix}^T , \qquad
f(z,t,y,h) = \begin{pmatrix} f_p(x_p,t) - h \\ f_m(s_m,\lambda,t,y,h) \\ f_a(x_m,y,t) \end{pmatrix}^T , \qquad (4.5)
$$

and the set $F(z,t) = \{f(z,t,y,h) \,|\, y \in Y(t), \ h \in P(t)\}$ of which f is the so-called
selector. Then we obtain the following generalized (contingent) equation for the
composite system

$$
\dot{z} \in F(z,t) \tag{4.6}
$$

with $\{(4.1), (4.3), (4.4)\}$ being the selecting set of equations.
Obviously, (4.6) could enclose far more noise than that of y without changes to our
further discussion, but extensions may obscure our point which is the use of the
method.

Given y,h, the solution to (4.6) through $(z^o, t_o) = (x_o^o, x_m^o, \lambda^o, t_o)$
$\in \Delta_o^2 \times \Lambda_o \times R = \Delta_o^3 \times R$ will now be designated by $z(z^o, t_o, \cdot) : R \to R_+^{2N} \times R^\ell$ and
the family of such solutions by $Z(z^o, t_o)$. Each solution is represented by the
solution curve $z(z^o, t_o, J_o) = \{z(z^o, t_o, t) \,|\, t \in J_o\}$ in $R_+^{2N} \times R^\ell \times R$. The set
$Z(z^o, t_o, t) = \{z(z^o, t_o, t) \,|\, z(\cdot) \in Z(z^o, t_o)\}$ in $R_+^{2N} \times R^\ell$ is the attainable set from
(z^o, t_o) at t and $Z(z^o, t_o, J_o) = \{Z(z^o, t_o, t) \,|\, t \in J_o\}$ in $R_+^{2N} \times R^\ell \times R$ is the flow
(or reachable set) during J_o from (z^o, t_o). We want the family $Z(z^o, t_o)$ on some
specified $\Delta_o^3 \times R$. Sufficient conditions imposed on our system to achieve the
latter do not narrow the problem in practical terms, see Filippov (1971) for
details.

We shall now specify the objectives. The identifier will be said to stabilize
(4.6) on some Δ_o^3 if given $f_p(\cdot)$ and $Y(t)$, $t \in J_o$ there are $f_m(\cdot)$, $f_a(\cdot)$ and $P(\cdot)$
such that $(z^o, t_o) \in \Delta_o^3 \times R$ implies $Z(z^o, t_o, J_o) \subseteq \Delta_o^3$. The latter is called stabilizability set. The union of such sets in $R_+^{2N} \times R^\ell$ forms the region of stabilizability.

We designate $||\cdot||$ to be a norm in R^{2N}, R^{ℓ}. Given $\eta' > 0$, let

$M_{\eta'} = \{(x_p, x_m) \in R^{2N}_+ \mid ||x_m - x_p|| < \eta'\}$ be the η'-neighborhood of the diagnonal

set M in R^{2N}_+, cf. Section 2. Given Δ^2_o, let $M_{\eta'} = \Delta^2_o \cap M_{\eta'}$, see Figure 3a.

Further, given $\eta'' > 0$, let $M_{\eta''} = \{\lambda \in \Lambda_o \mid ||\lambda|| < \eta''\}$. Then we denote

$M_{\eta} = M_{\eta'} \times M_{\eta''}$, see Figure 3b.

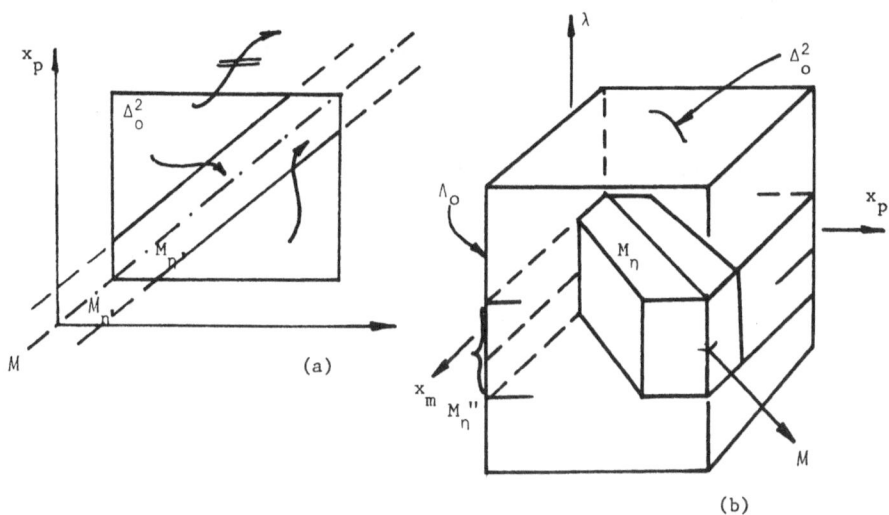

FIGURE 3. Converging solutions secure stabilization and identify the ecosystem

The identifier will be said to be η-<u>compatible</u> with the ecosystem on some Δ^3_o after

t_{η} if given $\eta', \eta'' > 0$, $T_{\eta} \geq 0$, $f_p(\cdot)$ and $Y(t)$, $t \in J_o$ there are $f_m(\cdot)$, $f_a(\cdot)$ and

$P(\cdot)$ such that $(z^o, t_o) \in \Delta^3_o \times R$ implies $Z(z^o, t_o, t) \subset M_{\eta}$ for all $t \geq t_{\eta} = t_o + T_{\eta}$.

M_{η} will be called the η-<u>compatibility</u> set. Defined as above, stabilizability and

compatibility are independent properties, but we shall seek them on the same Δ^3_o.

Let Δ_o^3 be bounded and $N(\partial\Delta_o^3)$ be a neighborhood of its boundary. We define

$\Delta_\varepsilon^3 = N(\partial\Delta_o^3) \cap \bar{\Delta}_o^3$ with $\text{int}\Delta_\varepsilon^3 = N(\partial\Delta_o^3) \cap \Delta_o^3$, to be considered the __safety-zone__, cf.

Leitmann-Skowronski (1977), about $\partial\Delta_o^3$ which shall guarantee the avoidance of the

latter by the solution curves from Δ_o^3.

__PROPOSITION 4.1.__ Given Δ_o, Λ_o, suppose there are $f_m(\cdot)$, $f_a(\cdot)$, $P(\cdot)$ and a

bounded function $V_s : R_+^{2N} \times R^\ell \times R \rightarrow R$, all defined on $\Delta_\varepsilon^3 \times R$ such that

(i) $V_s(z,t) < V_s(z^s,\tau)$ for all $(z,t) \in \text{int}\Delta_\varepsilon^3 \times R, (z^s,\tau) \in \partial\Delta^3 \times R$.

(ii) At every $t \in J_o$ we have $V_s(z,t) \leq V_s(z^o,t_o)$ for all

$\quad z \in Z(z^o,t_o,t)$, $(z^o,t_o) \in \text{int}\Delta_\varepsilon^3 \times R$.

Then the identifier stabilizes (4.6) on Δ_o^3.

__PROOF.__ Any solution curve from $\Delta_o^3 \times R$ must traverse $\Delta_\varepsilon^3 \times R$ in order to

leave $\Delta_o^3 \times R$. Hence it suffices to show that the solution curves from

$\Delta_\varepsilon^3 \times R$ do not leave $\Delta_o^3 \times R$. Let us suppose that there is $(z^o,t_o) \in \Delta_\varepsilon^3 \times R$

such that

$$Z(z^o,t_o,J_o) \cap \partial\Delta_o^3 \neq \phi \qquad (4.7)$$

Then there is $t_1 > t_o$ such that $Z(z^o,t_o,t_1) \cap \partial\Delta_o^3 \neq \phi$ and thus one can choose

$z^1 \in Z(z^o,t_o,t_1)$ such that by (i) we have $V_s(z^1,t_1) > V_s(z^o,t_o)$ which contradicts

(ii) yielding (4.7) false.   QED.

Note that the Proposition 4.1 does not require smoothness of V_s nor does it

prevent some increase in t. Note also that the contradiction in the proof is still

valid and thus the Proposition 4.1 still holds if the inequalities between the

values of V_s, strong in (i) and weak in (ii), are exchanged. It is easily observed

that for smooth V_s a stabilizability set Δ_o^3 will have to be defined by $\partial\Delta_o^3$ taken

as a level of V_s. A suggestion for a suitable Liapunov function based on practice

is to do the opposite - build V_s from its level defined by $\partial\Delta_o^3$. It may be useful

even if it is only a numerical construction, cf. Sticht-Vincent-Schultz (1975).

PROPOSITION 4.2. Assume that the identifier stabilizes (4.6) in Δ_o^3. Moreover, given the desired η', $\eta'' > 0$, $T_\eta \geq 0$ suppose that the corresponding $f_m(\cdot)$, $f_a(\cdot)$, $P(\cdot)$ as well as some additional smooth function $V_\eta : \overline{CM}_\eta \times R \to R$, $CM = \Delta_o^3 - M_\eta$, are such that

(i) $\nu_\beta \leq V_\eta(z,t) \leq \nu_\alpha$,

   where $\nu_\alpha = \sup V_\eta(z,t) \,|\, z \in \overline{CM}_\eta \cap \partial\Delta_o^3$, $t \in R$;

   $\nu_\beta = \inf V_\eta(z,t) \,|\, z \in \overline{CM}_\eta \cap \overline{M}_\eta$, $t \in R$.

(ii) $\dfrac{\partial V_\eta}{\partial t} + \nabla V_\eta\,(z,t) \cdot f(z,t,y,h) \leq -\dfrac{1}{T_\eta}\,(\nu_\alpha - \nu_\beta)$,

   for all $y \in Y$, $h \in P$, $t \in J_o$.

Then the identifier is also η-compatible on Δ_o^3 .

PROOF. The assumed stabilization means

$$(z^o,t_o) \in \Delta_o^3 \times R \Rightarrow Z(z^o,t_o,J_o) \subset \Delta_o^3 \ . \tag{4.8}$$

Now consider an arbitrary $(z^o,t_o) \in \overline{CM}_\eta \times R$ and an arbitrary solution $z(\cdot) \in Z(z^o,t_o)$ such that along the corresponding curve $z(z^o,t_o,\tau) \in \overline{CM}_\eta$, $\tau \in [t_o,t]$. From (ii) we have $\dot{V}_\eta(z(\tau),\tau) \leq -\dfrac{1}{T_\eta}\,(\nu_\alpha - \nu_\beta)$ along this curve. Integrating $\dot{V}_\eta = -\dfrac{1}{T_\eta}\,(\nu_\alpha - \nu_\beta)$ on $[t_o,t]$ we obtain the estimate

$$t \leq t_o + T_\eta \left(\frac{V_\eta(z^o,t_o) - V_\eta(z,t)}{\nu_\alpha - \nu_\beta} \right) \ . \tag{4.9}$$

By (i), $V_\eta(z,t) - \nu_\beta \geq 0$ and $V_\eta(z^o,t_o) - \nu_\alpha \leq 0$, whence $V_\eta(z^o,t_o) - V_\eta(z,t) \leq \nu_\alpha - \nu_\beta$ which substituted into (4.9) implies $t \leq t_o + T_\eta$ meaning that for $t > t_o + T_\eta$ the solution curve is no longer in $\overline{CM}_\eta \times R$. In view of (4.8) the curve must enter M_η. There is no return to $\overline{CM}_\eta$ at any later instant since then the conditions (i), (ii) contradict by the same argument as in the proof of the Proposition 4.1.   QED.

Presently the suggestion is that ∂M_η should define the Liapunov level.

For all practical purposes, at the present stage of development this Section is a suggestion only. Particularly, for eco-systems it has conjectural character requiring further work. Wider, control theoretic and more systematic study will be published (Leitmann and Skowronski) elsewhere. The state identification aspect may be also found in Skowronski (1980).

REFERENCES

Filippov, A.F. 1971. The Existence of Solutions of Generalized Differential Equations, Math. Notes (Mathematichiskie Zamietki), 10, 608-611.

Kudva, P. and Narendra, K.S. 1972. An Identification Procedure for Linear Multi-Variable Systems, Yale University, Boston Center Techn. Rep., CT-48.

Leitmann, G. and Skowronski, J. 1977. Avoidance Control, J. Optimiz. The. and Appl. 23, 581-591.

Narendra, K.S. and Kudva, P. 1974. Stable Adaptive Schemes for System Identification and Control, Part I, IEEE Trans, Systems-Man-Cybern, SMC-4, 542-551.

Skowronski, J.M. 1980. Strongly Nonlinear Observer Which Stabilizes under Uncertainty, Third IMA Conf. on Control Theory, Sheffield, UK.

Sticht, D.J., Vincent, T.L., and Schultz, D.G. 1975. Sufficiency Theorems for Target Capture, J. Optimiz. The. and Appl., 17, 523-543.

Yoshizawa, T. 1966. Stability Theory by Liapunov's Second Method, Math. Soc. of Japan Publ. #9, Tokyo.

COMPUTATIONAL DIFFICULTIES IN THE IDENTIFICATION AND OPTIMIZATION OF CONTROL SYSTEMS

H. T. Banks[†]
Lefschetz Center for Dynamical Systems
Division of Applied Mathematics
Brown University
Providence, Rhode Island   02912

As more realistic models for resource management are developed, the need for efficient computational techniques for parameter estimation and optimal control involving nonlinear vector systems will grow. We discuss some of the difficulties associated with such computational schemes and also report on results available for identification and control of several classes of systems which are of increasing importance in a number of areas of applications.

INTRODUCTION

In this presentation, we consider computational methods for <u>vector</u> dynamical systems of the type

$$\dot{x}(t) = f(t,x(t),u(t),\beta), \qquad 0 \le t \le T , \tag{1}$$

$$x(0) = x_0$$

or

$$x(i+1) = F(i,x(i),u(i),\beta), \qquad i = 0,1,2,\ldots,M - 1 , \tag{2}$$

$$x(0) = x_0 ,$$

with $x = (x_1,\ldots,x_n) \in R^n$, $u = (u_1,\ldots,u_m) \in R^m$, and $\beta = (\beta_1,\ldots,\beta_\nu) \in R^\nu$. The methods discussed are for the related problems of parameter estimation (identification) and optimization (optimal control) in "state" models which are assumed to be based on accepted (or hypothesized) mechanisms for growth, competition, harvesting, etc. in resource management. As one considers increasingly realistic models that include multi-species effects, age structure, multiple trophic levels, etc., (e.g., see (May et al. 1979), (Clark 1976), (Clark 1976a)) the analytic techniques employed in early studies of scalar or two dimensional vector models (see (Clark 1976)) will

[†] This research was supported in part by the National Science Foundation under grant NSF-MCS 79-05774, in part by the Air Force Office of Scientific Research under AFOSR 76-3092C, and in part by the U.S. Army Research Office-Durham under grant AROD-DAAG-79-C-0161

undoubtedly prove indadequate and one should find computational procedures such as those outlined below quite useful even in investigations of <u>qualitative</u> features of the more sophisticated models.

Our discussions here will be divided into two main sections. First, we give a brief review of standard ideas and techniques for identification and optimization problems involving ordinary differential equation models such as (1). Least squares techniques entailing such standard iterative techniques as gradient and conjugate gradient procedures will be outlined for the parameter identification problem for (1). Brief mention of maximum likelihood estimator ideas will be made. We next turn to necessary conditions for optimal control problems governed by (1). Our emphasis here will be on the celebrated Pontryagin maximum principle and the resulting two point boundary value problem that must be solved.

The second part of our presentation will be devoted to an indication of recent results (the development of which was mostly motivated by needs in other areas of applications such as aerodynamics, biochemistry and radiation biology) for special systems that we feel should be of use in resource management studies. We discuss the difficulties associated with identification and control of models with delays (hereditary systems) and offer tested ideas to alleviate some of these difficulties. Finally, procedures for optimization problems with underlying discrete systems such as (2) will be briefly outlined. Special problems in which the time intervals Δt_i between the control actions u(i) are themselves part of the optimization parameters appear to be of interest in certain terrestrial management endeavors and these will be discussed.

PARAMETER IDENTIFICATION

Consider the equation (1) in which the vector function u represents input to the system, β is a vector of unknown parameters (e.g., time constants, growth rates, etc.) and we are able to observe the "output" vector $y(t) = Cx(t)$. Here $y = (y_1, y_2, \ldots, y_p) \in R^p$ and C is a known $p \times n$ matrix. In practice, we make observations $\{\hat{y}^1, \hat{y}^2, \ldots, \hat{y}^K\}$ of the output y at times $t_1 < t_2 < \ldots < t_K$ (with or without perturbing the system via the input u). If we denote by $x(\cdot;\beta)$ the corresponding solution of (1), $\hat{y}^i$ is then an observation for $y(t_i;\beta) = Cx(t_i;\beta)$ and a <u>Least Squares Estimate</u> (LSE) of β is a solution to the problem of choosing $\beta \in \mathcal{B} \subset R^\nu$ so as to minimize

$$J(\beta) = \sum_{i=1}^{K} |\hat{y}^i - y(t_i;\beta)|^2 . \tag{3}$$

Here $\mathcal{B}$ is some given parameter set determined by physical, biological, economic, chemical, etc., restraints on the system.

For nonlinear vector systems, the problem of finding a $\beta^* \epsilon \, \mathcal{B}$ to minimize (3) is not, in general, solvable by analytic methods and some type of iterative numerical scheme is called for to produce estimates $\{\beta^j\}$ that hopefully converge to a "best fit" parameter β^*. Such a procedure can be formally stated as:

(i)   Guess an initial estimate β^0. (To make a good initial guess is important and one usually relies on his "knowledge" of the parameters although "apparent" values for parameters – which may have little relation to actual physical or biological limits – often play an important role in biological and ecological models!)

(ii)   Given an estimate β^j, generate a next estimate β^{j+1} by some formula $\beta^{j+1} = F(\beta^j)$. (For example, many popular methods are based on the iterate formula $\beta^{j+1} = \beta^j + \alpha_j d^j$ where d^j is a "direction" and α_j is a "step-size" parameter.)

(iii)   If $|\beta^{j+1} - \beta^j|$ is less than some given error tolerance, stop the procedure, accepting β^{j+1} as our value for β^*. Otherwise, set $\beta^j = \beta^{j+1}$ and return to step (ii) to generate a next estimate.

Among the most popular methods for the choice in (ii) above are the so called descent methods (see Chapter X of (Banks and Palatt 1975)) which include the gradient and conjugate-gradient techniques as special cases. Letting $\nabla J = \frac{\partial J}{\partial \beta}$ denote the gradient (partial derivative) of J with respect to β, the gradient method employs the iterative formula

$$\beta^{j+1} = \beta^j + \alpha_j [-\nabla J(\beta^j)]$$

where α_j is determined by a one-dimensional search on α in $\beta^j - \alpha \nabla J(\beta^j)$ to minimize J. The conjugate-gradient or conjugate-directions methods modify this choice of directions d^j (one no longer chooses $d^j = -\nabla J(\beta^j)$, but a choice related to this direction is made) in an attempt to insure that "independent" directions are chosen on subsequent steps in the algorithm. This is especially profitable in problems where narrow, steep "valleys" are present in the surface J as a function of the parameters. Among the more popular conjugate directions methods are those of Fletcher-Reeves and Daniel (Ortega and Rheinboldt 1970). In practice, hybrid methods combining gradient (G) and conjugate-gradient (CG) steps in some pattern (e.g., G, CG, CG, CG, G, CG, CG, CG, G, CG,...) often are found to perform better than an algorithm employing a single type of formula at each step.

A somewhat different approach to the least squares fit to data utilizes the method of quasilinearization. We shall not discuss these ideas here, but interested parties may consult (Banks and Groome 1973), (Bellman and Kalaba 1965), or Chapter X of (Banks and Palatt 1975).

An alternative to the LSE as formulated above is the <u>Maximum Likelihood Estimator</u> (MLE) for β. This procedure is based on the following considerations. In many problems, the observations $\{\hat{y}^i\}$ are corrupted by noise so we might write $\hat{y}^i = y(t_i;\beta^*) + Z_i$ where Z_i is the random measurement noise at time t_i and β^* is the true parameter value. If one assumes that $Z_1,\ldots,Z_K$ are independent random variables with identical probability density functions h (an assumption that is often <u>not</u> true in practice), then the joint probability density function is given by $\tilde{h}(z_1,z_2,\ldots,z_K) = \prod_{i=1}^{K} h(z_i)$. Loosely speaking, the function $\tilde{h}$ has its maximum at those values of $(z_1,\ldots,z_K)$ that are most likely to occur. Therefore, we might devise a procedure for estimating β^* on the basis that the observed values of $Z_i = \hat{y}^i - y(t_i;\beta^*)$ correspond to those that are most likely to occur, i.e.,

$\max \tilde{h}(z_1,\ldots,z_K) = \tilde{h}(Z_1,\ldots,Z_K)$. If we thus consider the function

$F(\beta) \equiv \tilde{h}(\hat{y}^1 - y(t_1;\beta),\ldots,\hat{y}^K - y(t_K;\beta))$, we might expect this function to be maximized at the value $\beta = \beta^*$. An MLE for β^* is then defined to be a value $\bar{\beta}$ which yields a maximum for $F(\beta)$. Instead of maximizing F, we define the <u>likelihood</u> function

$$L(\beta) = \ln F(\beta) = \sum_{i=1}^{K} \ln h(\hat{y}^i - y(t_i;\beta))$$

and equivalently seek a maximum for this function. This usually reduces to employing a procedure to determine $\bar{\beta}$ so that $\frac{\partial L}{\partial \beta}(\bar{\beta}) = 0$. A more detailed discussion of MLE's can be found in almost any standard text on statistics or estimation (Deutsch 1965), (Mood and Graybill 1963), (Cramer 1946), (Lindgren 1962).

We note that in certain cases the MLE reduces to the LSE (Deutsch 1965, p.136). For example, assume that Z is scalar, Gaussian with mean zero and variance σ^2. Then $h(z) = (1/\sqrt{2\pi}\,\sigma)\exp(-z^2/2\sigma^2)$ and

$$L(\beta) = \sum_{i=1}^{K} \ln \frac{e^{-(\hat{y}^i - y(t_i;\beta))^2/2\sigma^2}}{\sqrt{2\pi}\,\sigma}$$

$$= \sum_{i=1}^{K} \left[-\ln(\sqrt{2\pi}\,\sigma) - \frac{[\hat{y}^i - y(t_i;\beta)]^2}{2\sigma^2} \right]$$

$$= -K\ln(\sqrt{2\pi}\,\sigma) - \sum_{i=1}^{K} \frac{1}{2\sigma^2} [\hat{y}^i - y(t_i;\beta)]^2 \ .$$

Therefore maximizing L is equivalent to minimizing the least squares function J defined in (3).

OPTIMAL CONTROL

For this class of problems, we assume that the parameters β in (1) are known and define $g(t,x,u) \equiv f(t,x,u,\beta)$. In addition to the initial conditions in (1) we impose terminal or target conditions $x(T) \in T_1$ where T_1 is some given desired (smooth) subset of R^n. The control system is thus defined by

$$\dot{x}(t) = g(t,x(t),u(t)) , \qquad 0 \le t \le T ,$$

$$x(0) = x_0 \qquad\qquad\qquad\qquad\qquad (4)$$

$$x(T) \in T_1$$

and control functions $u: [0,T] \to U$, U a given restraint set, are to be chosen from a prescribed set U of admissible controls. The optimal control problem consists of choosing $u \in U$ so as to minimize

$$I(u) = g^0(x(T)) + \int_0^T f^0(s,x(s),u(s))ds$$

subject to (4). Here g^0, f^0 are given "payoff" or "cost" functions.

The necessary conditions for (x^*,u^*) to be a solution of this problem that we present here are the first order conditions that are analogues to the conditions $F'(\zeta) = 0$ employed in calculus when minimizing a scalar function F of one variable. More correctly, they are, roughly speaking, the function space analogue to the well-known Lagrange Multiplier Rule for constrained minimization in multivariable calculus problems (see Luenberger 1969), (Kirk 1970), (Hestenes 1975), (Bryson and Ho 1969)). Numerical computations are almost always essential for solving these necessary conditions in the case of vector systems that realistically model bio-logical or engineering phenomena. While the method of dynamic programming has re-ceived wide spread publicity, we shall not discuss it here since for vector systems of dimension greater than two it is often very difficult, if not impossible, to implement this method.

For $\psi_0 \in R^1$ and $\psi = (\psi_1, \ldots, \psi_n) \in R^n$, we define the "Hamiltonian" func-tion (scalar-valued) by

$$H(t,x,u,\psi_0,\psi) = \psi_0 f^0(t,x,u) + \psi \cdot g(t,x,u) \quad .$$

The Pontryagin Maximum Principle may then be stated (for a careful statement and proof, see (Berkovitz 1974) or (Fleming and Rishel 1975)) as follows:

PMP: If (x^*, u^*) is a solution to the above problem, there exist a nontrivial $n+1$ vector function $t \rightarrow (\lambda_0, \lambda(t)) = (\lambda_0, \lambda_1(t), \ldots, \lambda_n(t))$ such that

(a) λ_0 = constant ≤ 0 ,

$$\dot{\lambda}(t) = -\frac{\partial H}{\partial x}(t, x^*(t), u^*(t), \lambda_0, \lambda(t)) , \qquad 0 \leq t \leq T ,$$

(b) $\lambda(T)$ is orthogonal to T_1 at $x^*(T)$,

(c) $H(t, x^*(t), u^*(t), \lambda_0, \lambda(t)) = \max_{v \in U} H(t, x^*(t), v, \lambda_0, \lambda(t))$ for $0 \leq t \leq T$.

The boundary conditions on λ given in (b) are called transversality condi-tions and the condition in (c) is of course the "maximum condition" that the op-timal pair (x^*, u^*) must satisfy. The equations in (a), which can be equivalently written

$$\dot{\lambda}(t) = -\lambda_0 \frac{\partial f^0}{\partial x}(t, x^*(t), u^*(t)) - \lambda(t)\frac{\partial g}{\partial x}(t, x^*(t), u^*(t)) ,$$

are called the "costate" or "adjoint" or "multiplier" equations. These equations, with the boundary conditions from (b), are coupled with the system (4) and togeth-er they form a two-point boundary value problem (TPBVP) for a 2n-vector system (i.e., (4), (a), (b) must be solved simultaneously). Note, however, that these equations involve the (unknown but sought after) control function u^* which must be determined from the condition (c). But condition (c) involves the functions x^* and λ which are to be determined from (4), (a), and (b). Thus, conditions (a), (b), (c) taken with (4) constitute a TPBVP with a coupled maximization condition.

There is, in general, no hope of solving this analytically and some type of computational scheme is required.

Just as in the case of the identification problems discussed above, one can develop iterative schemes based on gradient, conjugate-gradient, etc. ideas to generate a sequence of pairs (x^j, u^j) that (hopefully) converge to (x^*, u^*). However, in this case the iterations are made in function space. The gradient of I with respect to u is an operator with kernel $\frac{\partial H}{\partial u}$; that is,

$$\nabla I(u; \delta u) = \int_0^T \frac{\partial H}{\partial u} \delta u \, dt$$

and the descent type procedures will thus use $\frac{\partial H}{\partial u}$ as the gradient "direction." For example, a procedure based on the gradient method is given by the steps:

[1] Choose an initial estimate u^0 ;

[2] Given the jth estimate u^j, generate the next estimate u^{j+1} by:

    (i) Use u^j in $\dot{x} = g(t, x, u)$, $x(0) = x_0$, to compute x^j.

    (ii) Use u^j, x^j with $\dot{\lambda} = \frac{-\partial H}{\partial x}(t, x, u, \lambda_0, \lambda)$, $\lambda(T) \perp T_1$ at $x^j(T)$ to compute λ^j .

    (iii) Compute $D^j = \frac{\partial H}{\partial u}(t, x^j, u^j, \lambda_0, \lambda^j)$.

    (iv) Put $u^{j+1} = u^j + \alpha_j D^j$ where α_j is determined by a one-dimensional search to minimize $I(u)$;

[3] If $u^{j+1} \approx u^j$, terminate the procedure, accepting u^{j+1} as the estimate for u^*; otherwise, set $u^j = u^{j+1}$ and return to [2] above .

There are many variations on the iterative procedure outlined here, some of which involve different choices for the directions D^j . For further discussions, one can consult (Kirk 1970), (Polak 1973), (Banks and Palatt 1975). An excellent survey of other methods can also be found in (Polak 1973).

SYSTEMS WITH DELAYS

Most investigations of resource management are, by necessity, based on models for "population" behavior (resource increase, utilization, etc.). Due to the complexity in ecological chains, it is often desirable to consider multiple trophic levels (e.g., vegetation-herbivore-carnivore systems) in the models used. Other factors that play an important role are age structure, environment recovery, and delay in recruitment. All of these considerations lead, in any realistic modeling effort (May et al. 1979, p. 275), (Cushing 1977), (Clark 1976), (May 1973, p. 73), (May 1976, p. 6) to systems with delays. Such systems have played an increasingly important role in engineering and other areas of applications and in recent years efforts to develop mathematical techniques to aid in investigating delay systems have grown. We report on some of the recent results obtained from these efforts in hopes that these techniques will also prove of value to investigators of models for renewable resources.

Consider for the moment the delay system

$$\dot{x}(t) = f(t,x(t),x(t-r),u(t),\beta) , \qquad 0 \le t \le T$$

$$x(\theta) = \phi(\theta) , \qquad -r \le \theta \le 0 , \tag{5}$$

which, even though we include only a discrete delay term, is an infinite dimension-al "state" system (similar to a partial differential equation). This poses im-mediate added difficulties for the identification and control problems discussed in the sections above. However, in the case of certain parameter estimation prob-lems, there are even more serious questions that must be entertained. Often, in addition to the parameters β in (5), one also wishes to estimate the delay r so that the parameter identification problem consists of estimating $\gamma = (\beta,r)$ or, in the case of multiple delay systems, $\gamma = (\beta,r_1,\ldots,r_\mu)$. A moment's reflection (see (3) and the related discussions) will reveal that in order to use the standard techniques discussed above, one needs to be able to compute $\frac{\partial x}{\partial \beta}(t_i;\beta)$ and, in this case, $\frac{\partial x}{\partial r}(t_i;r)$. However, this derivative does not, in general, even exist! Consider the following example:

$$\dot{x}(t) = x(t-r), \qquad t > 0 ,$$

$$x(\theta) = \begin{cases} 0 & -1 \le \theta \le 0 \\ 1 & \theta < -1 . \end{cases}$$

For $r \le 1$, the solution is given by $x(t;r) = 0$ for $t \ge 0$. For $r > 1$, say $r = 1 + \varepsilon$, we find $x(t;r) = t$ for $0 \le t \le \varepsilon$, and $x(t;r) = \varepsilon$ for $\varepsilon \le t \le 1 + \varepsilon$. It is then very easy to show that for $0 < t_i < 1$, $\frac{\partial x}{\partial r}(t_i; 1)$ does not exist. The techniques we summarize here will overcome this difficulty as well as those pre-sented by the infinite-dimensional state aspects of (5).

We sketch the ideas for the simplest linear delay system. Results are avail-able for the most general linear systems, including those with distributed delay terms (e.g., $\int_{t-r}^{t} k(\theta)x(\theta)d\theta$) - see (Banks and Burns 1978), (Banks et al. 1979),

(Banks and Kappel 1979), (Banks et al. 1979b) - as well as for a rather general class of nonlinear system (Banks 1979), (Banks 1980).

Consider the simplest linear delay system

$$\dot{x}(t) = A_0(\beta)x(t) + A_1(\beta)x(t-r) + Bu(t) \qquad t > 0 ,$$

$$x(\theta) = \phi(\theta) , \qquad -r \le \theta \le 0 , \qquad\qquad (6)$$

$$y(t) = Cx(t) .$$

Here $x \in R^n$ and A_0, A_1 are n×n matrix functions that are assumed to depend on the parameters β in a smooth manner. Let $\gamma = (\beta, r)$ with Γ some given admissible parameter set. We wish to choose $\gamma \in \Gamma$ so as to minimize

$$J(\gamma) = \sum_{i=1}^{K} |\hat{y}^i - y(t_i;\gamma)|^2$$

where the $\hat{y}^i$ are observations for $y(t_i;\gamma) = Cx(t_i;\gamma)$. The methods we propose entail approximating (6) by high order differential equations in which the parameters γ appear smoothly so that we may apply standard techniques. That is, we have an approximating identification problem:

Find $\gamma^N \in \Gamma$ so as to minimize

$$J^N(\gamma) = \sum_{i=1}^{K} |\hat{y}^i - C^N w^N(t_i;\gamma)|^2$$

where

$$\dot{w}^N(t) = A^N(\beta,r)w^N(t) + B^N u(t)$$

$$w^N(0) = w_0^N(\phi) . \qquad\qquad (7)$$

Here (7) is an approximating system for (6) that is finite dimensional ($w^N \in R^{\rho(N,n)}$) with the dimension ρ depending on the index of approximation N and the dimension n of the original system (6). One can, for the schemes we have developed, argue that γ^N converges (as $N \to \infty$) rapidly to a solution γ^* of the original estimation problem. For details see (Banks et al. 1979b). We shall not discuss this here since the arguments are quite technical. Essentially, one treats (6) as as abstract system in a Hilbert space and employs classical Ritz type ideas: Approximate the problem on finite-dimensional subspaces. Convergence properties and error estimates can then be obtained using approximation results from linear semigroup theory.

These approximation ideas (for both the parameter estimation and optimal control problems) have been developed to date for two specific classes of schemes:

(I) The "averaging" approximations (Banks and Burns 1978): These approximations are based on approximations of states $\psi \in L_2^n(-r,0)$ by step functions $\psi^N = \sum_{j=1}^{N} a_j \chi_j$ where χ_j is the characteristic function for $I_j^N = [\frac{-jr}{N}, \frac{-(j-1)r}{N}]$, $j = 1,2,\ldots,N$, and a_j is the integral average of ψ over the subinterval I_j^N; i.e.,

$$a_j = \frac{N}{r} \int_{I_j^N} \psi(s)\,ds \quad .$$

The matrix $A^N(\beta,r)$ in (7) turns out to be very simple to use in computations, being given by

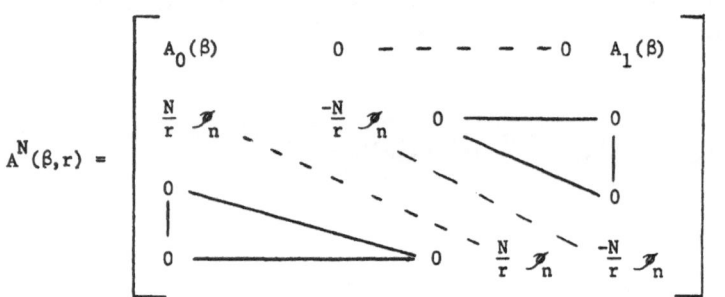

Here $\mathcal{I}_n$ is the $n \times n$ identity matrix.

(II) Spline based approximations (Banks and Kappel 1979): These approximations are based on best L_2 approximations of states in subspaces consisting of spans of standard spline basis elements. The theory has been developed for arbitrary order splines and has been tested numerically for piecewise linear and cubic elements. The matrix A^N in these cases is not as simple as in the case of the averaging approximations, but the methods still are quite easily implemented.

We have carried out extensive testing of these methods as applied to identification and optimization problems. These findings (Banks et al. 1979a) indicate
that our theoretical estimates are supported by computational efficacy of the methods. We present here a sample of the numerical findings in a parameter estimation
problem. The example is typical of tubular column reactor problems discussed in
(Banks et al. 1979b) in which one designs experiments to determine a transport coefficient β for the product of an enzyme catalyzed reaction taking place in the
column. Enzyme pellets are packed in the column and it is also desired to estimate
the "diffusion delay" r associated with diffusion from the interior of the pellet
to the exterior flow region of the column. Mathematically, we wish to estimate β
and r in the system

$$\dot{z}_1(t) = E_0(\beta)z_1(t) + E_1(\beta)z_1(t-r) + G$$

$$0 \leq t \leq 8 ,$$

$$\dot{z}_1(t) = E_0(\beta)z_i(t) + E_1(\beta)z_i(t-r) + E_2 z_{i-1}(t-2)$$

$$i = 2,3,4 ,$$

$$z_1(\theta) = (1,0)^T$$

$$-2 \leq \theta \leq 0 ,$$

$$z_i(\theta) = (0,0)^T$$

where

$$E_0(\beta) = \begin{bmatrix} -3-\beta & 0 \\ 2\beta & 0 \end{bmatrix} \quad E_1(\beta) = \begin{bmatrix} 0 & \beta \\ 0 & -2\beta \end{bmatrix} \quad E_2 = \begin{bmatrix} 3 & 0 \\ 0 & 0 \end{bmatrix} \quad G = \begin{bmatrix} 1 \\ 0 \end{bmatrix} .$$

In this example we have each vector z is 2-dimensional so that the example is an
8 vector system $x = (z_1,z_2,z_3,z_4)^T$. It is assumed that only the first component of
z_4 can be observed. Data was generated using the "true" values $\beta^* = -3$ and $r^* = 2$.
A least squares criterion as in (3) was used along with a standard iterative package (an IMSL package). The approximate problem for (7) with $N = 24$ was investigated for both the averaging and linear spline schemes. Start-up values in the iter-
ation were $\gamma^0 = (\beta^0, r^0) = (-4,3)$. For the averaging approximations, satisfactory
convergence was not obtained, while the iterates for the spline scheme are given in
Table 1. (This illustrates dramatically one difference between theory and practice-
in theory both schemes should produce good approximations-only the spline scheme
does in this example.)

TABLE 1

β	r
-4.0	3.0
-3.717	2.341
-3.535	1.954
-3.436	1.929
-3.172	1.903
-3.109	1.921
-3.069	1.950
-3.039	1.987
-3.004	1.994

Use of these approximation ideas in optimization problems as described above has also been investigated both theoretically and numerically. Briefly, if one wishes to minimize I(u) (as defined above) over U subject to (6), one again approximates by a problem for a high dimension ordinary differential equation (7). One then seeks to find a $u^N \varepsilon U$ that minimizes

$$I^N(u) = g_N^0(w^N(T)) + \int_0^T f_N^0(s, w^N(s), u(s)) ds$$

subject to (7). Here g_N^0, f_N^0 are appropriately chosen approximations for g^0, f^0. Again, one can argue (see Banks and Burns 1978, Banks and Kappel 1979, Banks et al. 1979) that solutions (x^N, u^N) of the approximating problems converge as $N \to \infty$ to (x^*, u^*), a solution of the original problem, for a large class of payoff functions g^0, f^0. Extensive numerical results can be found in (Banks et al. 1979a), while theoretical results for nonlinear systems can be found in (Banks 1979, Banks 1980).

DISCRETE SYSTEMS

Parameter estimation and optimal control problems for discrete systems (2) can be formulated in much the same way as indicated above for the continuous systems (1). We outline briefly the optimization problem and results, because we feel that these have direct applications in resource management and because the associated necessary conditions are not simple analogues to those for the continuous system problem in the case of many nonlinear systems.

Assuming that the parameter β is known, we seek a set of values u = {u(0), u(1),...,u(M-1)} from a given constraint set $U \subset R^m$ so as to minimize

$$I(u) = g^0(x(M)) + \sum_{i=0}^{M-1} f^0(i,x(i),u(i))$$

subject to (2). The necessary conditions (see PMP above) are again defined in terms of a Hamiltonian function

$$H_i(v) = f^0(i,x(i),v) + \lambda(i+1)^{\cdot}F(i,x(i),v,\beta) , \qquad i = 0,1,\ldots,M-1 ,$$

and an adjoint or costate equation

$$\lambda(i) = -\frac{\partial H_i}{\partial x(i)} (u^*(i)) , \qquad i = 1,\ldots,M-1 , \tag{8}$$

with appropriate boundary conditions (see (Almquist and Banks 1976) for a more precise statement). However, the global maximum condition (the analogue of (c) in PMP above)

$$H_i(u^*(i)) = \max_{v \in U} H_i(v)$$

is valid for nonlinear system problems only under very strong convexity conditions on F and U (which are not usually satisfied in applications). Rather there is a local maximum condition

$$\frac{\partial H_i}{\partial u(i)} (u^*(i)) \cdot [u^*(i)-v] \geq 0 , \qquad v \in U \tag{9}$$

or, in some cases,

$$\frac{\partial H_i}{\partial u(i)} (u^*(i)) = 0 , \tag{10}$$

that is valid for most nonlinear system problems of interest. For a precise statement of the global and local conditions one can consult Chapter VII, Section 4 of (Neustadt 1976). The last chapter (Notes and Historical Comments) of (Neustadt 1976) contains an account of the interesting historical development (marked with some confusion and incorrect claims) of necessary conditions for discrete systems control problems along with a rather complete bibliography. Computational techniques based on (9) and (10) that are modifications of the iterative schemes (e.g., descent methods) discussed above have been developed and employed on a wide variety of discrete system problems. For examples see (Almquist and Banks 1976), (Bryson and Ho 1969), (Kirk 1970).

In the usual applications of discrete system theory in engineering and in some biological models - see (Almquist and Banks 1976), (Clark 1976a) - one has an underlying continuous time process in which "discontinuous" events or phenomena (e.g., treatments, harvests, etc.) occur. That is, one has a process evolving in time with major events (controls $u(i)$) occurring at time intervals $\Delta t_0, \Delta t_1, \ldots, \Delta t_{M-1}$ apart as depicted in Figure 1.

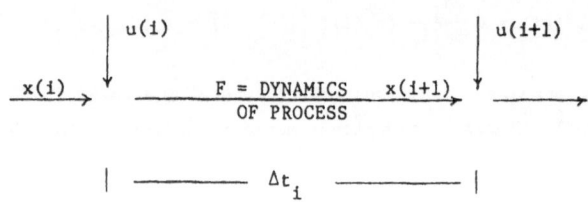

FIGURE 1. Discrete Modeling For A Continuous Process

In the optimization problem formulated above, it is assumed that Δt_i, the time between control actions $u(i)$ and $u(i+1)$, is a fixed parameter. However, in our efforts to extend the work of (Almquist and Banks 1976) on fractionated therapy for tumors, we have been obliged to consider optimization problems for discrete systems in which the times Δt_i are additional parameters over which one wishes to optimize. That is, in addition to dealing with variable <u>levels</u> of discrete control actions, one wishes to allow variable times between the control actions. We have recently tested computational ideas such as those described in (Almquist and Banks 1976) for problems where one seeks optimal "controls" $w(i) = (u(i), \Delta t_i)$ for problems with systems of the form $x(i+1) = G(i,x(i),w(i))$. Our preliminary findings (details will appear in a manuscript in the future) suggest that efficient methods can be readily developed for discrete time variable-control-time problems via careful modification and extension of existing schemes.

We believe that such techniques could be of use in resource management studies in which one considers variable times between harvesting, etc. For example, they could be important in forestry management problems since it has been observed (Aber et al. 1978, Aber et al. 1979) that rotation length (time between control

actions) is perhaps in some cases a more important parameter than harvesting intensity (level of control action). Simulation and optimization studies of various
cutting regimes must, due to the complexity of the underlying ecological models
(Botkin et al. 1972), ultimately involve computational procedures for vector models.

ACKNOWLEDGEMENT

I am grateful to Dan Botkin for fruitful conversations which led me to include
the comments on discrete systems in my presentation at the Workshop. I would also
like to acknowledge the substantial contributions made by P. Daniel in development
of the software packages used in numerical experiments for the methods discussed in
my presentation.

REFERENCES

Aber, J.D., Botkin, D.B., and Melillo, J.M. 1978. Predicting the Effects of Different Harvesting Regimes on Forest Floor Dynamics in Northern Hardwoods, Can. J.
Forest Res., 8, 306-315

Aber, J.D., Botkin, D.B., and Melillo, J.M. 1979. Predicting the Effects of Different Harvesting Regimes on Productivity and Yield in Northern Hardwoods, Can. J.
Forest Res., 9, 10-14

Almquist, K.J. and Banks, H.T. 1976. A Theoretical and Computational Method for
Determining Optimal Treatment Schedules in Fractionated Radiation Therapy, Math.
Biosci., 29, 159-179

Banks, H.T. 1979. Approximation of Nonlinear Functional Differential Equation
Control Systems, J. Optimization Theory Appl., 29, 383-408

Banks, H.T. 1980. Identification of Nonlinear Delay Systems Using Spline Methods,
Proc. International Conf. on Nonlinear Phenomena in the Math. Sciences, (U. Tex.
Arlington, June 16-20, 1980), Academic Press, to appear.

Banks, H.T. and Burns, J.A. 1978. Hereditary Control Problems: Numerical Methods
Based on Averaging Approximations, SIAM J. Control and Optimization, 16, 169-208

Banks, H.T., Burns, J.A., and Cliff, E.M. 1979. Spline-based Approximation Methods
for Control and Identification of Hereditary Systems, in International Symposium on
Systems Optimization and Analysis, A. Bensoussan and J.L. Lions, eds., Lecture
Notes in Control and Info. Sci., Vol. 14, Springer, Heidelberg, 314-320

Banks, H.T., Burns, J.A., and Cliff, E.M. 1979a. A Comparison of Numerical Methods
for Identification and Optimization Problems Involving Control Systems with Delays,
Brown University LCDS Tech. Rep. 79-7, Providence

Banks, H.T., Burns, J.A., and Cliff, E.M. 1979b. Parameter Estimation and Identification for Systems with Delays, SIAM, J. Control and Optimization, submitted

Banks, H.T. and Groome, G.M., Jr. 1973. Convergence Theorems for Parameter Estimation by Quasilinearization, J. Math. Anal. Appl., 42, 91-109

Banks, H.T. and Kappel, F. 1979. Spline Approximations for Functional Differential Equations, J. Diff. Eq., 34, 496-522

Banks, H.T. and Palatt, P.J. 1975. Mathematical Modeling in the Biological Sciences, Brown University LCDS Lec. Notes 75-1, Providence

Berkovitz, L.D. 1974. Optimal Control Theory, Springer-Verlag, New York

Botkin, D.B., Janak, J.F., and Wallis, J.R. 1972. Some Ecological Consequences of a Computer Model of Forest Growth, J. Ecology, 60, 849-872

Bryson, A.E. and Ho, Y.C. 1969. Applied Optimal Control, Blaisdell Publ., Waltham

Bellman, R. and Kalaba, R. 1965. Quasilinearization and Nonlinear Boundary-Value Problems, American Elsevier, New York

Clark, C.W. 1976. Mathematical Bioeconomics, John Wiley, New York

Clark, D.W. 1976a. A Delayed Recruitment Model of Population Dynamics with an Application to Baleen Whale Populations, J. Math. Biol., 3, 381-391

Cramér, H. 1946. Mathematical Methods of Statistics, Princeton Univ. Press, Princeton

Cushing, J.M. 1977. Integrodifferential Equations and Delay Models in Population Dynamics, Springer Lecture Notes in Biomath., 20, New York

Deutsch, R. 1965. Estimation Theory, Prentice-Hall, Englewood Cliffs

Fleming, W.H. and Rishel, R.W. 1975. Deterministic and Stochastic Optimal Control, Springer-Verlag, New York

Hestenes, M.R. 1975. Optimization Theory, John Wiley, New York

Kirk, D.E. 1970. Optimal Control Theory, Prentice-Hall, Englewood Cliffs

Lindgren, B.W. 1962. Statistical Theory, Macmillan, New York

Luenberger, D.G. 1969. Optimization by Vector Space Methods, John Wiley, New York

May, R.M. 1973. Stability and Complexity in Model Ecosystems, Princeton Univ. Press, Princeton

May, R.M. 1976. Theoretical Ecology: Principles and Applications, Blackwell, Oxford

May, R.M., Beddington, J.R., Clark, C.W., Holt, S.J., and Laws, R.M. 1979. Management of Multispecies Fisheries, Science, 205, 267-277

Mood, A.M. and Graybill, F.A. 1963. Introduction to the Theory of Statistics, McGraw-Hill, New York

Neustadt, L.W. 1976. Optimization, Princeton Univ. Press, Princeton

Ortega, J.M. and Rheinboldt, W.C. 1970. Iterative Solution of Nonlinear Equations in Several Variables, Academic Press, New York

Polak, E. 1973. An Historical Survey of Computational Methods in Optimal Control, SIAM Rev., 15, 553-584

ECONOMIC MODELS OF FISHERY MANAGEMENT

Colin W. Clark
Mathematics Department
University of British Columbia
Vancouver, B.C., V6T 1W5

This paper briefly surveys the existing literature on economic models
of commercial fisheries, including: optimization models, models of
competitive fishing, and models of fishery regulation. An appendix
discusses economic aspects of Southern Ocean resource exploitation.

INTRODUCTION

Fisheries must be managed. This is by now a truism, the demonstration of
which lies in the many examples of depleted fish stocks and overexpanded fishing
fleets throughout the world. What still remain controversial are (a) the appropriate objectives of management (Holt and Talbot, 1977; May et al. 1979), and
(b) the appropriate means of achieving these objectives (Pearse, 1979a; FAO, 1980).

With few exceptions, fish stocks are not the property of any individual or
firm. Until recently few marine fish stocks were even claimed by any individual
state, the main exception being anadromous species. The recently proclaimed 200-
mile zones have altered the situation considerably, however. Since 1954 at least
(see Gordon, 1954), economists have attributed the depletion of fisheries to the
lack of ownership, or property rights in fish stocks (see also Cheung, 1970).
According to the Gordon theory, exploitation of such a common-property resource[*]
will intensify up to the level of zero profit, i.e., until each exploiter is making
zero profits (relative to his other economic opportunities). For some fish stocks -
namely highly valued species that are easily captured - this zero-profit level may
correspond to a biologically depleted population, and in theory even to an extinct
population.

From the economic point of view, the zero-profit solution is an inefficient
one (i.e., not a Pareto, or welfare optimum), regardless of the biological outcome (see Clark, 1976, Ch. 2). By controlling the level of exploitation it would
be possible in principle to achieve positive profits (more precisely, positive

---

[*] The same rule applies to any common property resource, renewable or otherwise.

economic rents) from the fishery. If these rents were distributed to the fishermen, the fishermen would be better off. In a sense, the fishermen would then become property owners, and would enjoy the wealth that normally accrues to such
persons. Alternatively, the state could retain ownership of the resource, charging
fishermen for the right to catch fish. Intermediate positions - profit (rent) shar-
ing - are also conceivable.

The proviso "in principle" in the preceding paragraph is a serious one, however.
The establishment of property rights always involves certain costs - enforcement
costs, transaction costs, etc. Unless the potential benefits of establishing property rights exceed these costs, it is not worthwhile to establish a system of
rights. In the case of marine resources especially, there is also the fundamental
problem of jurisdiction. Obviously property rights cannot be established unless
the resource lies within the jurisdiction of some authority capable of enforcing
those rights. The notorious difficulties of managing international fisheries are
indubitably a basic consequence of the lack of well defined jurisdiction over these
resources.

Fish populations that remain within the 200-mile zone of a single state can now
be considered as the property of that state. But it may not be obvious what further rights delineations, if any, are necessary in order to guarantee the efficient
operation of the fishery. Indeed, it may not be entirely clear what "efficient"
signifies, especially in a dynamic, intertemporal sense. Finally, from the practical point of view the economic implications of various conceivable management
policies may not be obvious.

All these questions can be addressed by the application of control-theoretic
methods to appropriate mathematical models of the fishery. First, the ideal economic optimum can be characterized by modelling a centralized, profit (or rent, or
welfare) maximizing agency - the "sole owner" of the fishery. Such a model - first
discussed by A. D. Scott (1955) - is useful to the discussion, even if sole ownership is usually a practical impossibility. Secondly, the purely competitive fishery can be modelled by assuming that individual fishermen (or vessel owners) attempt to maximize their net revenue flow, taking the current state of the fish stock
as given - and adjusting their input of fishing effort in response to stock abundance, market prices, etc. Because of the externality inherent in the common-
property resource, the competitive model possesses a different solution from the
sole-ownership model, as first explained by H.S. Gordon (1954).

A third, but much more difficult class of optimization models deals with
strategic, or game-theoretic behavior. Such models seem appropriate for fisheries
exploited by a fairly small (and restricted) number of vessels (Clark, 1979), or
for fisheries dominated by a small number of purchasing-processing companies (Clark
and Munro, 1980). Game theoretic models also seem needed for the study of trans-

boundary and other international fisheries (Munro, 1979; Vincent, 1980). Various alternative behavioral hypotheses can be incorporated into game-theoretic models, and much work remains to be done in these investigations. Finally, by suitable modification, fishery models can be constructed with the aim of predicting the response of the fishery to alternative forms of regulation (Clark, 1980).

To date most fishery models incorporating any degree of economic realism have employed highly simplified biological submodels (single species, general production, deterministic). For practical purposes the development and analysis of more complex biological models is important. Unfortunately, a fundamental difficulty arises from the paucity of biological data, implying that biologically realistic models cannot usually be adequately validated. Gross uncertainty is a uniform characteristic of fishery management, it seems.

What role do models (and optimization analysis) play in such a situation? Some scientists express the view that models are only useful when they "exactly" represent the real world. (One biologist recently said to me, "Oh, you'd never be able to model this fishery - it's far too complex.") The older school of fishery economists has asserted that dynamic optimization models are largely of academic interest, because of the lack (and unavailability) of necessary data (Crutchfield, 1979, p. 743; Scott, 1979, p. 727).

An alternative viewpoint is that models are especially useful in complex situations where data is minimal. It is only by constructing and studying models (whether verbal, experimental, mathematical, or computer models) that we can hope to increase our understanding of such systems. Indeed, if we want to predict the consequences of interference with some natural system, there is no alternative but first to model the system. Our predictions may not be very accurate, but to quote N. Keyfitz, "no model, no understanding" (Keyfitz, 1977).

The remainder of this paper describes, very briefly, the historical development of mathematical models pertaining to the economics of fisheries. For further details, and for all mathematical calculations, the reader is referred to the quoted literature. Finally in a brief appendix, I discuss some economic aspects of Southern Ocean fisheries.

THE "COMMON-PROPERTY" FISHERY

H. S. Gordon (1954) was the first economist to present an analysis of the common-property fishery, although some biologists had earlier published similar theories (e.g., Graham, 1952). Gordon's discussion is non-mathematical, and considers only equilibrium situations. The Gordon theory, however, is readily amalgamated with the dynamic fishery model developed at about the same time by

M. B. Schaefer (1954). The equations of the Gordon-Schaefer model are:

$$\frac{dx}{dt} = F(x) - h(t) \tag{1}$$

$$h(t) = qE(t)x(t) \tag{2}$$

$$\pi(t) = ph(t) - cE(t) \tag{3}$$

where: x denotes the fish population biomass, F(x) denotes the natural growth rate, h(t) denotes the catch rate, E(t) denotes fishing effort, q is a constant (the "catchability" coefficient), π(t) denotes net revenue flow, p is the price of fish, and c is the cost of fishing effort. In the original Schaefer model, F(x) is quadratic (logistic):

$$F(x) = rx(1 - x/K); \tag{4}$$

this is easily generalized, e.g., to general production models of the type used by Pella and Tomlinson (1969).

Gordon's model is obtained by setting dx/dt = 0 in the above. One easily derives the corresponding relationship between effort E and total (equilibrium) revenue ph, viz.

$$TR = pqKE(1 - qE/r) \ .$$

Figure 1 shows this "yield" curve, together with the total cost curve, TC = cE .

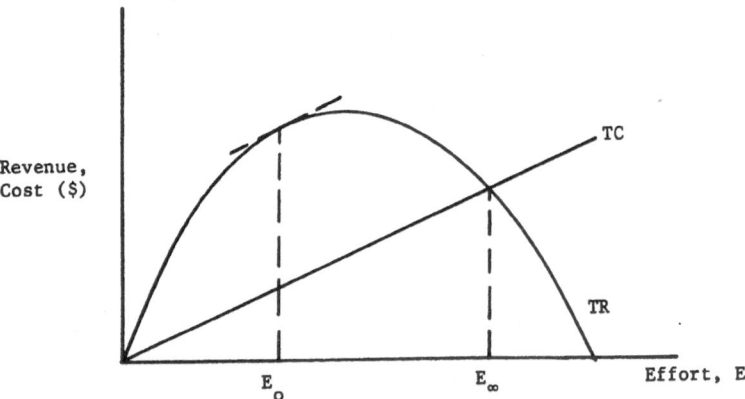

FIGURE 1. The Gordon Model

Now comes the main prediction of the model: in an open-access common-property fishery, effort will expand to an equilibrium level $E = E_\infty$ at which net revenue equals zero. From the model equations, this implies

$$x = x_\infty = \frac{c}{pq}, \qquad E = E_\infty = \frac{F(x_\infty)}{qx_\infty}. \tag{5}$$

Gordon calls this the "bionomic equilibrium" of the common-property fishery. If the parameter values are such that the curves TR, TC are located as in Fig. 1, this equilibrium (x_∞, E_∞) corresponds to depletion of the fish stock: $x_\infty < K/2$, with sustainable yield $F(x_\infty) < F(K/2) = \max F(x)$. Clearly (says Gordon) this is an economically inefficient result.

Gordon's model was extended by V. L. Smith (1969), who added an equation of the form

$$\frac{dE}{dt} = \alpha\pi = \alpha(pqx(t) - c)E(t) \tag{6}$$

to describe the response of effort E to the observed revenue flows π. The plane autonomous system of Eqs. (1) and (6) has a unique stable equilibrium corresponding to Gordon' bionomic equilibrium (5). However, the system may have complex eigenvalues at (x_∞, E_∞), in which case the convergence will be oscillatory. In any case, there are likely to be large "overshoots" of the bionomic equilibrium solution. The Smith model has been fitted to data from the North Pacific fur seal fishery in the early 20th century by J. Wilen (1976).

A further extension of the model has been achieved by L. G. Anderson (1977), who investigates the behavior of individual fishermen or vessels under the explicit assumption of pure competition in the capture of fish. A model of this sort seems essential for formulating predictions regarding the effects of various regulatory policies (Clark, 1980 ; see below). To describe Anderson's model, assume for simplicity a collection of identical fishing vessels, each capable of exerting a level of effort $E_i \geq 0$. Adopt a standard cost function $c(E_i)$ for effort, with U-shaped marginal cost (Figure 2.) The revenue (or "profit") maximizing hypothesis is then that each vessel is operated so as to maximize net revenue flow:

$$\underset{E_i \geq 0}{\text{maximize}} \quad \pi_i = pqxE_i - c(E_i). \tag{7}$$

The solution is clearly (see Figure 2):

$$c'(E_i) = pqx \qquad \text{if} \quad pqx > r$$

$$E_i = 0 \qquad \text{if} \quad pqx < r. \tag{8}$$

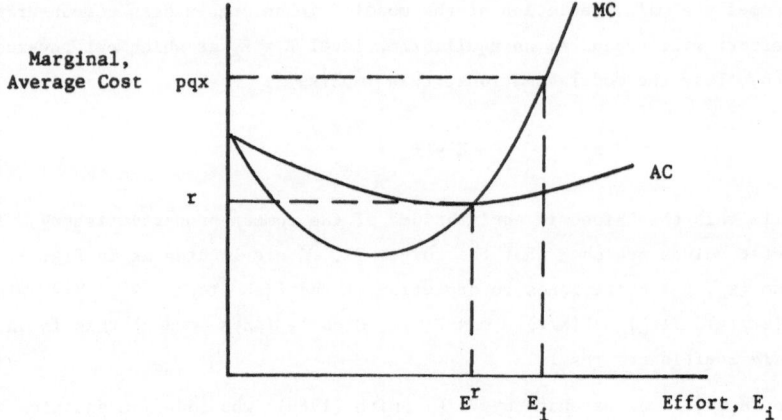

FIGURE 2.   Effort cost curves, $MC = c'(E_i)$ and $AC = c(E_i)/E_i$

With N vessels participating, total effort equals

$$E_{tot} = NE_i(x)$$

where $E_i(x)$ is given by Eq. (8). Bionomic equilibrium is achieved when net revenue per vessel vanishes, i.e., when

$$x = x_\infty = \frac{r}{pq} \, , \quad E_{tot} = N_\infty E_r = \frac{F(x_\infty)}{qx_\infty} \, . \tag{9}$$

This agrees with Eq. (5) in the case that $c(E) = cE$.

   Modelling the dynamics associated with this solution is somewhat problematical. If N is assumed fixed (unrealistic for the open-access case!), total effort is given by

$$E_{tot}(x) = NE_i(x)$$

which is a discontinuous function of x, giving a somewhat disturbing knife-edge aspect to the equilibrium at x_∞ . (If $N < N_\infty$ then equilibrium is reached at $x > x_\infty$, with each vessel earning positive net revenue. This case, which contradicts the open-access hypothesis, is discussed further below.) The knife-edge effect is reduced if we suppose that different vessels possess different degrees of efficiency, i.e., different cost curves $c_i(E_i)$. The i-th vessel will then drop out of the fishery when $x < x_{i\infty} = r_i/pq$. Ultimately only the most efficient vessels will survive (or at any rate, inefficient vessels will be replaced by efficient ones). But this brings us back to the knife-edge - and in fact suggests that it is a real phenomenon.

None of the above models separates variable costs from fixed costs. The decision to enter a fishery often involves a significant investment in vessel and gear. This investment is likely to be risky, especially in an open-access fishery. To a certain extent, the decision will involve strategic elements (how many other vessels will enter?), and concepts from the theory of (dynamic) games thus become relevant (Clark, 1979). But, these complexities aside, it is important to consider the irreversible nature of investment decisions in fisheries. A fishing vessel has few uses other than fishing, and very little value as scrap (which was nevertheless the fate of many of the large pelagic whaling factory ships in the 1960's). Basically, a new vessel will enter the fishery only if expected net revenues will provide an appropriate rate of return on investment, i.e., provided

$$\int_0^T \pi_i(t) e^{-\delta t} \, dt \geq FC_o \tag{10}$$

where the integral represents discounted expected returns over the life of the vessel, at discount rate δ, and FC_o = fixed (capital) cost of initial investment. On the other hand, once constructed the vessel will tend to remain in the fishery provided that revenues exceed variable costs, i.e., provided $\pi_i(t) > 0$, or equivalently, $x > x_m$. We thus conclude that there exists a "gap" in the biomass levels separating the exit decision from the entry decision for new vessels. The reader is referred to Clark (1979, 1980) and Clark et al. (1979) for further analysis of this difficult question. An unpublished paper of Eswaran and Wilen (1978) is also relevant.

OPTIMIZATION ANALYSIS

The foregoing models all support the conclusion that the unregulated open-access fishery is unsatisfactory. The original Gordon analysis, for example, shows that (with the configuration of curves shown in Fig. 1) both biological and economic yields could be increased, over the long run, by a decrease in fishing effort. Gordon's paper identifies E_o (Fig. 1) as the optimum economic position, where TR - TC is maximized. Quite a bit of fuss was made when it was noted that this economic "optimum" was somewhat _more_ conservative (less effort, greater biomass) than the traditional MSY solution favored by the biologists (Gordon, 1955; Crutchfield and Zellner, 1962). However, when one leaves the equilibrium manifold to which Gordon's analysis is restricted, the validity of this conclusion fails.

The first attempt to inject dynamics into the Gordon model was due to A. D. Scott (1955), who analyzed (verbally) the behavior of a "sole owner" of the fishery. This theory seems to have been first expressed mathematically by Crutchfield and Zellner (1962), who formulated the following variational problem

$$\underset{\{E(t)\}}{\text{maximize}} \quad \int_0^\infty e^{-\delta t} \pi(x,E) \, dt \tag{11}$$

subject to:
$$\frac{dx}{dt} = F(x) - qxE , \quad x(0) = x_o \tag{12}$$

$$x \geq 0, \quad E \geq 0 . \tag{13}$$

The simplest case arises when π is linear in E, as in Eq. (3) above and as in the original Gordon model (Clark and Munro, 1975). In this case, there exists a singular solution $x = x^* = $ constant, which is determined by the conditions

$$F'(x) + \frac{\partial \pi/\partial x}{\partial \pi/\partial h} = \delta$$

$$F(x) = h \tag{14}$$

or, explicitly from $\pi = (p - c/qx)h$:

$$F'(x^*) + \frac{cF(x^*)}{qx^*(pqx^*-c)} = \delta . \tag{15}$$

The optimal effort policy is then

$$E = \begin{cases} E_{max} & \text{when } x > x^* \\ E^* = F(x^*)/qx^* & \text{when } x = x^* \\ 0 & \text{when } x < x^* . \end{cases} \tag{16}$$

The reader is referred to the above papers for discussion of the economic interpretation of these results, and for various extensions.

It is easy to show that:

$$x_\infty < x^* < x_o \tag{17}$$

(where $x_o = \frac{1}{2}(K+x_\infty)$ is the optimum biomass level resulting from the Gordon optimization model). Thus the "sole owner" of the fishery achieves an equilibrium with a larger fish stock (and less effort) than the open-access fishery, but because of the positive discount rate, with less fish (and more effort) than Gordon's static optimum. This analysis indicates, among other things, that there exists a "cost of conservation" resulting from the discounting of future revenues. While the social justification for such discounting may be open to question (Page, 1977), there is no doubt that the sole owner (and the fishing industry in general) will exhibit some degree of time preference. Whether this will have a serious effect on the optimal biomass x^* depends on the ration between r and δ, the intrinsic growth rate and the rate of discount. Species with low intrinsic growth rates, such as whales (Clark, 1975), will be especially vulnerable to depletion because of this time-preference effect. But even for species with high growth potential, the analysis suggests that the fishing industry may be somewhat less enthusiastic about conservation than the static Gordon "optimal economic yield" would indicate - even ignoring the game-theoretic implications of this solution (Vincent, 1980).

Nevertheless, it remains clear that for most fish species, the open-access situation is unsatisfactory both to the fishing industry and to society as a whole. The next question is, how can a more satisfactory solution be accomplished? Observation of managed fisheries suggests that this question may be more difficult than it seems.

MODELS OF FISHERY REGULATION

In spite of the extensive literature on fishery economics, which was developed over two and a half decades following Gordon's paper, few studies were addressed explicitly to the theory of fishery regulation (Scott, 1979, p. 725). Recently, however, in response to the establishment of 200-mile zones, economists and others have vigorously taken up this question (Anderson, 1980; Pearse, 1979a,b; Clark, 1980; Lewis and Matthews, 1979). Because of the many possibilities, modelling soon tends to become complex, and only the first steps can be described here.

The main management alternatives are:

i   total catch quotas, or equivalent controls
ii license limitation
iii fiscal interference (i.e., taxes on catch or on effort)
iv allocated fishing rights (i.e., quotas on catch or effort).

The continuous-time model described above is a bit too simplistic to allow for a comparative analysis of all these alternatives. Total catch quotas, for example, are usually specified on an annual basis, and the fishery is closed as soon as the total logged catch reaches the predetermined quota; this method has been used in the management of fisheries for tropical tuna, Pacific halibut, anchoveta, whales, and other species. This situation can be conveniently modelled by means of a discrete-time (seasonal) analog of our basic model. But since the details are rather tedious, I shall refer the reader to my paper (Clark, 1980) for further discussion. The main prediction of the analysis is the following.

Suppose that in the absence of regulation the fishery is (or would be) over-exploited biologically, i.e., $x_\infty < x_{MSY}$. Annual catches are then regulated so as to keep the fish stock at the estimated MSY level (for example). At this stock level, catch rates per vessel are higher than at the unregulated equilibrium x_∞, as are net revenues. If the number of vessels is not restricted, it follows as in the Gordon theory that additional vessels will enter the fishery, driving net revenues again to zero. In other words, fishery regulation based on total catch quotas alone leads to an _increase_ in the capacity of the fishing fleet, beyond the unregulated level - even though this level of capacity is itself excessive. Experience with many quota-regulated fisheries agrees with this prediction.

Similar predictions can be established for other traditional methods of fishery regulation. Mesh-size regulations, for example, designed to prevent the capture of small fish, will result in an increase in the population biomass, and hence (as above) to increased vessel revenues, which ultimately attract additional vessels. While these techniques do succeed, at least in principle, in preventing the depletion of fish stocks, from the economic viewpoint they are largely self-defeating. In some actual instances, capacity has increased to two or three times the necessary level, and indeed there is no theoretical upper limit to overcapacity. Under such circumstances, control of the fishery becomes increasingly difficult.

The next logical step would seem to involve a limitation of entry to the fishery, for example by licensing only a specific number N of fishing vessels. If we continue to assume that each vessel maximizes its net revenue flow π_i, we have as in Eq. (8)

$$c_i'(E_i) = pqx \quad (\text{unless } E_i = 0) \ . \tag{18}$$

It is easy to see that, regardless of N, this solution cannot be optimal, since the optimal solution is specified by the problem

$$\max_{(E_i)} \int_0^\infty e^{-\delta t} \sum_{i=1}^N \pi_i \, dt \tag{19}$$

subject to

$$\frac{dx}{dt} = F(x) - qx \sum_1^N E_i \ . \tag{20}$$

For this problem we obtain

$$c_i'(E_i) = (p-\lambda) \; qx \tag{21}$$

where λ is a positive "shadow price" (adjoint variable). Comparing (21) with (18), we conclude that the profit maximizing vessels all exert effort levels in excess of the optimum. The explanation is of course obvious - the vessels are still competing for the catch, and nothing in the cost structure forces one fisherman to account for the effects that his own catch has on the future catch rates of the other vessels. This analysis remains valid even when license limitation is used in conjunction with total catch regulation - as is currently the case, for example, in the regulation of Canada's Pacific salmon fisheries (Fraser, 1979).

Examination of Eq. (21) suggests one way in which "external" costs of fishing may be brought to bear on individual fishermen, namely by imposing a catch tax ("royalty") equal to the value of the shadow price λ . Net vessel revenue after taxes then becomes

$$\tilde{\pi}_i = (p-\lambda) \; qxE_i - c_i(E_i) \tag{22}$$

so that profit maximization now leads to the correct input of effort by each vessel. The same result could be achieved (for our model!) by taxing effort at the rate $qx\lambda$.

The use of taxes to correct for externalities is often recommended by economists on general principles. Taxes are thought to be relatively free from undesirable distortions in their effects on individuals' decisions. Also, the operation of a tax scheme is considered to be fairly simple and straightforward. However, in the case of fisheries severe problems arise, not the least of which is the political pressure that fishermen are likely to exert. Taxes of course have distributional as well as efficiency implications, and it is obviously the former that will be of greatest concern to the fishermen.

It is interesting to note, therefore, that an alternative management technique exists which has similar (in fact at our level of abstraction, identical) efficiency implications, but the opposite distributional implications, to taxes. This is the technique of allocated fishermen (or vessel) quotas, which are also assumed to be transferable. To model this technique, suppose that the i-th vessel has a quota Q_i on its catch rate:

$$h_i \leq Q_i \; , \tag{23}$$

and let $Q_T = \sum_{i-1}^{N} Q_i$ denote the total of these allocated quotas.

The quota Q_i may be smaller, or larger than what is desired by the operator of the i-th vessel. Thus some vessels may have unused quotas while other vessels wish to increase their quotas. Since quotas are assumed transferable, we may also imagine that a quota <u>market</u> is established, at which quota units are bought and sold, at some price m. (The costs of operating the market are ignored here.) Writing

$$\pi_i(x, h_i) = p h_i - c_i(h_i/qx) \tag{24}$$

we see that vessel i will wish to <u>buy</u> an additional quota unit, at price m, if and only if

$$\frac{\partial \pi_i}{\partial h_i}(x, Q_i) > m \tag{25}$$

where Q_i denotes the vessel's current quota holding. The equation $\partial \pi_i / \partial h_i = m$, or

$$c_i'(Q_i/qx) = (p-m)\ qx \tag{26}$$

thus determines the i-th vessel's demand schedule for quota units, as a function of the quota price m. It is easy to check that this demand curve has the expected negative slope. The total demand for quotas is obtained as usual by (horizontal) summation; see Fig. 3. Since the total quota is fixed at Q_T, an equilibrium price m is determined, as are the final quota holdings Q_i of each vessel. (Since x, and possibly Q_T, vary over time, both m and Q_i may also be time dependent.)

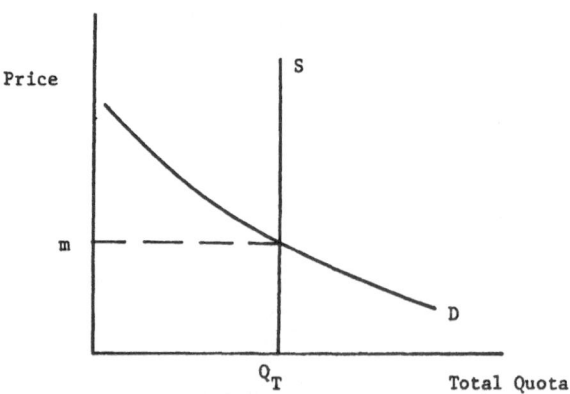

FIGURE 3. Supply and demand for allocated quotas

Now a rational vessel operator will either utilize his full quota or sell the excess:

$$h_i = Q_i = qxE_i \quad . \tag{27}$$

Hence Eq. (26) becomes

$$c_i'(E_i) = (p-m) \; qx \; , \tag{28}$$

which is identical in form with Eq. (21). The total quota Q_T can then be fixed so that

$$m = \lambda \quad . \tag{29}$$

In other words, the allocated quota system (with optimized total quota Q_T) is equivalent to a system of catch taxes, and either can be used to optimize the fishery economically. The quota price ("scarcity value" of quotas) is equated by the quota market to the shadow price, which equals the optimal tax.

Similarly, transferable effort quotas are equivalent to effort taxes, both being merely transformations of catch quotas or taxes.

So much for the model. What about the real world? The equivalences just established are quite obviously model-dependent, and a great deal of further analysis is required to elucidate the practical pros and cons of each method, not to mention possible combinations. The role of uncertainty, which I have ignored completely, promises to be most important. (See Anderson, 1980, or Pearse, 1979a,b, for further discussion.)

Let us note, to conclude this paper, that allocated catch quotas and catch taxes could easily be used in conjunction with each other. If τ represents the tax rate, and Q_T is optimal, we obviously get

$$m + \tau = \lambda \quad . \tag{30}$$

In this case, the proceeds of the fishery accrue jointly to the initial quota holders (via the quota market at price m), and the taxing authority. By choosing $\tau \in [0,\lambda]$, any desired distribution of benefits can be achieved.

REFERENCES

Anderson, L.G. 1977. The Economics of Fisheries Management. The Johns Hopkins University Press, Baltimore

Anderson, L.G. 1980. A comparison of limited entry fisheries management schemes, in United Nations FAO Fisheries Report No. 236, Rome: 47-74

Cheung, S.N.S. 1970. The structure of a contract and the theory of a nonexclusive resource. J. Law Economics 13: 49-70

Clark, C.W. 1974. Possible effects of schooling on the dynamics of exploited fish populations. J. Cons. Int. Expl. Mer 36: 7-14

Clark, C.W. 1975. The economics of whaling: a two-species model. In G.S. Innis (ed.) New Directions in the Analysis of Ecological Systems, Simulation Council Proc. Ser. 5(1): 111-119

Clark, C.W. 1976. Mathematical Bioeconomics: The Optimal Management of Renewable Resources. Wiley-Interscience, New York

Clark, C.W. 1979. Restricted access to common-property fishery resources: a game-theoretic analysis, in P.T. Lui (ed.), Dynamic Optimization and Mathematical Economics, Plenum, New York: 117-132

Clark, C.W. 1980. Towards a predictive model for the economic regulation of commercial fisheries. Can. J. Fish. Aqua. Sci. 37(7): 1111-1129

Clark, C.W., Clarke, F.H., and Munro, G.R. 1979. The optimal exploitation of renewable resource stocks: problems of irreversible investment. Econometrica 47: 25-49

Clark, C.W. and Munro, G.R. 1975. The economics of fishing and modern capital theory: a simplified approach. J. Envir. Econ. Manag. 2: 92-106

Clark, C.W. and Munro, G.R. 1980. Fisheries and the processing sector: some implications for management policy. Bell J. Econ. (In press)

Crutchfield, J.A. 1979. Economic and social implications of themain policy alternatives for controlling fishing effort. J. Fish. Res. Board Canada 36: 742-752

Crutchfield, J.A. and Zellner, A. 1962. Economic aspects of the Pacific halibut fishery. Fish. Indust. Res. 1

Eswaran, M. and Wilen, J.E. 1977. Expectations and adjustment in open-access resource use. Univ. of B.C. Dept. of Economics Resources Paper No. 10

Food and Agriculture Organization of the United Nations. 1980. Final Report of the ACMRR Working Party on the Scientific Basis of Management, FAO Fisheries Report No. 236, Rome

Fraser, G.A. 1979. Limited entry: experience of the British Columbia salmon fishery. J. Fish. Res. Board Canada 36: 754-763

Gordon, H.S. 1954. The economic theory of a common property resource: the fishery. J. Polit. Econ. 62: 124-142

Gordon, H.S. 1958. Economics and the conservation question. J. Law Economics 1: 110-121

Graham, M. 1952. Overfishing and optimum fishing. Cons. Perm. Intern. Expl. Mer: Rapp. et Proc.-Verb. 132: 72-78

Gulland, J.A. 1976. The stability of fish stocks. J. Cons. Int. Expl. Mer 37: 199-204

Holt, S.J. and Talbot, L.M. 1978. New principles for the conservation of wild living resources. Wildlife Monographs. No. 59.

Keyfitz, N. 1977. Applied Mathematical Demography, Wiley-Interscience, New York

Lewis, T.R. and Matthews, S.A. 1978. A certificate-leasing program for marine resource development. Unpublished m/s. Envir. Qual. Lab., Cal. Inst. Tech., Pasadena, Calif.

May, R.M., Beddington, J.R., Clark, C.W., Holt, S.J., and Laws, R.M. 1979. Management of multispecies fisheries. Science 205: 267-277

Munro, G.R. 1979. The optimal management of transboundary renewable resources. Can. J. Econ. 12: 354-376

Page, T. 1977. Conservation and Economic Efficiency. The Johns Hopkins University Press, Baltimore

Pearse, P.H. 1979a, ed. Symposium on Policies for Economic Rationalization of Commercial Fisheries. J. Fish. Res. Board Canada 36: 711-866

Pearse, P.H. 1979b. Property rights and the regulation of commercial fisheries. Univ. of B.C. Dept. of Economics Resources Paper No. 42

Pella, J.J. and Tomlinson, P.K. 1969. A generalized stock-production model. Bull. Inter-Amer. Trop. Tuna Comm. 13(3): 421-496

Schaefer, M.B. 1954. Some aspects of the dynamics of populations important to the management of commercial marine fisheries. Bull. Inter-Amer. Trop. Tuna Comm. 1:25-56

Scott, A.D. 1955. The fishery: the objectives of sole ownership. J. Polit. Econ. 63: 116-124

Scott, A.D. 1979. Development of economic theory on fisheries regulation. J. Fish. Res. Board Canada 36: 725-741

Smith, V.L. 1969. On models of commercial fishing. J. Polit. Econ. 77: 181-198

Vincent, T.L. 1980. Vulnerability of a Prey-Predator model under harvesting. (This Volume)

Wilen, J. 1976. Common-property resources and the dynamics of overexploitation: the case of the North Pacific fur seal. Univ. of B.C. Dept. of Economics Resources Paper No. 3.

APPENDIX

ECONOMIC ASPECTS OF RENEWABLE RESOURCE EXPLOITATION
IN THE SOUTHERN OCEAN[*]

Historically, the only really important fisheries in the Southern Ocean have been the whale fisheries. As everyone knows, the major whale species (particularly blue, fin, humpback) were severely depleted during the present century. With modern technology and the high demand for whale oil - and later, whale meat - Antarctic whaling was extremely profitable while the stocks lasted.

The traditional explanation for the depletion of the Antarctic whale stocks involves the competition between whalers, i.e., the Gordon model. But I have argued (e.g., Clark, 1975) that the whalers may actually have deliberately chosen a policy of severe depletion, since whale populations are a pretty poor "investment." Indeed, most of the decisions taken by the International Whaling Commission can be shown to agree, at least qualitatively (and to as close a fit as I can make, quantitatively) with the preferences of a profit maximizing industry. Even the current moratorium agrees with the theory (Clark, Clarke, and Munro, 1979). It will be interesting to observe future developments.

The possibility of a significant fishery on Antarctic krill, the main food source for whales, is now under discussion (see May et al. 1979). Such a development seems likely if and only if krill fishing becomes profitable - or seems so. This is true even under socialist economics - a krill fishery will be seriously considered only if productivity in that industry equals or exceeds productivity in alternative protein supply industries. At present there is no question that krill can be harvested in large quantities, and transformed either into an edible product, or into some form of fish meal. Indeed, small quantities of krill could probably be marketed at quite a high price, as dried "shrimp." The problem seems to be that the market for this particular product is very small - orders of magnitude below the estimated MSY of krill (100 - 1000 million tons). Thus most of the catch would have to be utilized either as fish meal or protein supplement, both of which currently demand low prices.

It seems unlikely, therefore, that a huge krill fishery is really imminent. Current increases in oil prices make the prospect seem even less likely.

---

[*] This appendix was written after the conference and was, to a certain extent, inspired by the workshop forums.

If, nevertheless, a substantial krill fishery does develop, the difficult and interesting question of joint management of krill and whales will arise. Here there is the possibility of a real conflict between long- and short-term benefits, just as in the exploitation of the whales themselves. That is, even though it might be clear that the sustainable yield from whales far outweighed in value the sustainable yield from krill (both not being attainable simultaneously), commercial interests would favor "krill now" over "whales later." It might even be argued that feeding the starving masses was more important than saving the whales (for later use) - a difficult argument to counter! It seems unlikely, however, that deliberate destruction of the krill resource would be desired: presumably krill is r-selected, not K-selected. Thus there should be a strong incentive to manage krill at least. The reported preference of Southern Ocean Treaty powers for a policy of "MSY for each species" could indicate a desire ultimately to harvest krill without having to worry too much about the ecological complexities of the Antarctic system.

Finally, what about the possibility (eventually) of an unexpected krill "crash"? The economic content of this question relates mostly to the extent to which krill concentrate into easily scooped up patches. In fact, it seems likely that a krill fishery (probably like the whales) would depend on such concentrations. This means not only that the fishery would be closely competing with whale populations feeding on krill, but also that the likelihood of a "bifurcation" in the krill-whale-fishery dynamics is increased (Clark, 1974, Gulland, 1977). This problem seems worthy of further study.

VULNERABILITY OF A PREY-PREDATOR MODEL
UNDER HARVESTING

Thomas L. Vincent
University of Arizona
Tucson, Arizona 85721, USA

Given a limit on the intensity of harvesting, the vulnerability of a
renewable resource system is taken to be a measure of how close the
resource can be driven to extinction under an arbitrary harvesting
schedule. The primary objective of this analysis is to demonstrate
how the concept of vulnerability can be used in the development of a
practical method for setting a limit on the intensity of harvesting.
Necessary conditions associated with controlling a system in the
boundary of a reachable set are employed to determine the vulnerability of a system directly.

Using a two species prey-predator model, it is shown that an estimate
of vulnerability for a system based on a constant harvesting analysis
may greatly underestimate the true vulnerability of the system. This
is simply because a non-constant harvesting program coupled with the
dynamical nature of the system can be used to amplify the vulnerability
effect.

INTRODUCTION

Simultaneous control of both species in the Lotka-Volterra prey-predator system model has been previously investigated by Goh, et al. (1974), Vincent, et al.
(1974), Vincent (1975), and others. More recently May, et al. (1979) using a new
model, somewhat akin to the Lotka-Volterra system, examined the consequences of
simultaneous constant effort harvesting of both the prey (Krill) and the predators
(Baleen Whales). Fishing efforts were restricted to values which would yield
positive equilibrium populations for both species. Within these limits, the model
yields stable equilibrium solutions under any constant fishing effort. The problem
of choosing a fishing effort for each fishery to maximize yield at equilibrium was
addressed and it was clearly demonstrated that the concept of maximum sustainable
yield (MSY) used in single species models is not directly applicable to the multi-
species situation.

Beddington and Cooke (1979) examined the same model under constant rate harvesting and demonstrated the instability (or regions of limited stability) for many
of the possible equilibrium solutions. The instability of constant rate harvesting
in some models has been previously noted by Goh (1977), and Beddington and May
(1977).

Analysis of the simultaneous harvesting of a multispecies system model involves
the sorting out of many interrelated questions. It is assumed here that a given

model does indeed reflect reality and that each species is harvested by a separate group of harvesters, (e.g., capelin fishermen and sealers, krill fishermen and whalers, etc.). Attention is then focused on the question of how to manage the system so as to "maximize" the harvested yield for each species without endangering any of the species in the exploited ecosystem. Neither of the populations is to be driven to unacceptably low levels.

Suppose that the manager of a multispecies prey-predator ecosystem is able to specify only an upper limit to the intensity of harvesting for each species. Under this supposition, the harvesting program (within these limits) actually employed by the harvesters represents a detail over which the manager has no control. The time sequence details of specific harvesting programs is not considered here. Rather, the analysis addresses the question: Given a specific model for a prey-predator ecosystem and the above specification of the managers input, what limits should the manager set for the intensity of harvesting, for both the prey and predator, so that each harvester can operate efficiently and, in addition, so that none of the species will be in danger of extinction no matter what harvesting programs (within the limits) are actually used by the harvesters?

For a given dynamical model for a prey-predator system, suppose that all possible positive equilibrium solutions under specified constant harvesting are determined. For each equilibrium solution, there corresponds a sustainable yield for both the prey and predator. The particular equilibrium solution which corresponds to the "maximization" of these yields is a problem in continuous static game theory (Vincent and Grantham, 1980). There are several solution concepts which are applicable to multispecies harvesting. Only the Nash solution concept is used here. At a Nash solution point the yield of a given harvester can not be increased by making a unilateral change in the harvesting effort by that harvester. The Nash concept applied to a single species ecosystem is equivalent to the maximum sustainable yield concept. A more complete analysis of the game theoretic aspects of this problem will appear elsewhere.

Any solution concept can be used to select a tentative limit for the intensity of harvesting. In particular, if the constant harvesting Nash solution results in population levels sufficiently high so that none of the species are considered endangered, then this solution is taken as a tentative limit for the intensity of harvesting. This limit remains tentative until the dynamical aspects of the system are properly assessed. That is, any tentative solution for harvesting limits obtained by assuming constant harvesting must not allow any of the species to become endangered under an arbitrary harvesting program (e.g., non-constant), satisfying the tentative limits.

Clearly if the tentative solution is dynamically unstable with respect to a constant harvesting program, then it is generally unsatisfactory. The exception would be that of a limit cycle sufficiently "close" to the equilibrium point that the endangered species criteria is still satisfied. Even if the tentative solution is dynamically stable with respect to a constant harvesting program, it may be possible to endanger one or both of the species with a non-constant harvesting program. One can check the tentative solution under a non-constant harvesting program by using the vulnerability method of Goh (1976) based on global Liapunov stability analysis (LaSalle and Lefschetz, 1961) or the vulnerability method of Vincent and Anderson (1979) which is based on controllable set theory (Grantham and Vincent, 1975). The latter method is used in this paper. The vulnerability analysis demonstrates that tentative solutions obtained from the constant harvesting program analysis will often not satisfy an endangered species requirement so that subsequent adjustment to the harvesting levels are necessary.

THE MODEL

Analysis is restricted to the simultaneous harvesting of a simple prey-predator system. In particular, the model parameters are chosen to simulate a Krill (prey) and Baleen Whale (predator) ecosystem. The following model is due to May, et al. (1979).

Let N_1 represent the prey population and N_2 represent the predator population. The dynamics of the system are given by

$$\dot{N}_1/N_1 = r_1(1 - \frac{N_1}{K}) - aN_2 - r_1 u_1 H_1$$

$$\dot{N}_2/N_2 = r_2(1 - \frac{N_2}{\alpha N_1}) - r_2 u_2 H_2$$

where r_1 is the intrinsic growth rate of the prey, u_1 is the prey harvest effort, if $H_1 = 1$ and is prey harvest rate if $H_1 = 1/N_1$, r_2 is the intrinsic growth of the predator, u_2 is the predator harvest effort if $H_2 = 1$ and is predator harvest rate if $H_2 = 1/N_2$. The dot refers to differentiation with respect to time. The model includes three additional constants, K (carrying capacity), a, and α. Note that with $N_2 = 0$ and $u_1 = 0$, $\lim_{K\to\infty} \dot{N}_1/N_1 = r_1$ and with $u_2 = 0$, $\lim_{N_1\to\infty} \dot{N}_2/N_2 = r_2$. In addition note that $u_1 H_1 = 1$ or $u_2 H_2 = 1$ corresponds to a per capita fishing rate at the intrinsic growth rate.

It is convenient to nondimensionalize the model. Let

$$x_1 = N_1/K$$

$$x_2 = N_2 a/r_1$$

$\gamma = r_1/\alpha a K$ (=$1/\nu$ in the May, et al. 1979, paper)

$$\tau = t r_1 \quad (t = \text{time})$$

$$\beta = r_2/r_1 \quad .$$

In terms of these nondimensional variables the model becomes

$$\overset{\circ}{x}_1/x_1 = 1 - x_1 - x_2 - u_1 H_1 \tag{1}$$

$$\overset{\circ}{x}_2/x_2 = \beta(1 - \gamma x_2/x_1 - u_2 H_2) \tag{2}$$

where $^\circ$ denotes differentiation with respect to nondimensional time.

Under constant harvesting the yield (for unit nondimensional time) y_1 of the prey species is given by

$$y_1 = u_1 x_1 H_1 \tag{3}$$

and the yield (per unit nondimensional time) y_2 of the predator species is given by

$$y_2 = \beta u_2 x_2 H_2 \tag{4}$$

EQUILIBRIUM SOLUTIONS (ISOCLINES)

Tentative harvesting limits are set by assuming constant harvest levels and examining the resulting positive equilibrium solutions (i.e., $\overset{\circ}{x}_1/x_1 = \overset{\circ}{x}_2/x_2 = 0$).

Case 1. $H_1 = H_2 = 1$ (Effort Harvesting)

At equilibrium, equations (1) and (2) reduce to

$$x_2 = 1 - x_1 - u_1 \tag{5}$$

$$x_2 = x_1(1 - u_2)/\gamma \quad . \tag{6}$$

For given harvesting efforts, the first equation yields points for which the rate of growth of prey are zero (prey isoclines) and the second equation yields points for which the rate of growth of predators are zero (predator isocline). The intersection of these two isoclines represents an equilibrium solution and is given by

$$x_1 = \gamma(1 - u_1)/(\gamma + 1 - u_2) \tag{7}$$

$$x_2 = (1 - u_1)(1 - u_2)/(\gamma + 1 - u_2) \quad . \tag{8}$$

All possible positive equilibrium solutions are obtained by choosing $0 \le u_1 \le 1$ and $0 \le u_2 \le 1$. For example with $\gamma = 1$, all possible equilibrium solutions are contained in the triangle of Figure 1 composed of the regions E, 5, 6, and 7. As pointed out by May et al. (1979), $\gamma = 1$ corresponds to the predators harvesting the prey at MSY. Assuming this to be a reasonable point of departure, all examples in this paper will use $\gamma = 1$.

The tentative constant harvesting limit is obtained by finding an equilibrium point which is a Nash solution for the yields. By substituting (7) and (8) into (3) and (4) the expressions for yield become

$$y_1 = \gamma u_1(1 - u_1)/(\gamma + 1 - u_2) \tag{9}$$

$$y_2 = \gamma \beta u_2(1 - u_1)(1 - u_2)/(\gamma + 1 - u_2) \quad . \tag{10}$$

For given system constants γ and β, the yields are functions of u_1 and u_2 only. For a range of values for γ, a unique Nash solution exists on the interior of the control set defined by $0 \le u_1 \le 1$, $0 \le u_2 \le 1$. The solution is obtained by setting $\partial y_1/\partial u_1 = \partial y_2/\partial u_2 = 0$ and is given by

$$u_1 = 1/2 \tag{11}$$

$$u_2 = (\gamma + 1) - \sqrt{\gamma(\gamma + 1)} \tag{12}$$

For $\gamma = 1$, the Nash solution for the control is given by

$$u_1 = .5, \ u_2 = .586 \tag{13}$$

with the equilibrium point located at

$$x_1 = .3536, \ x_2 = .1464 \tag{14}$$

with the corresponding yields

$$y_1 = .1768, \ y_2 = .0858 \tag{15}$$

Isoclines corresponding to the Nash control ($u_1 = .5$, $u_2 = .586$) are also drawn in Figure 1. These isoclines along with the original zero harvest ($u_1 = u_2 = 0$) isoclines divides the positive orthant into the eight regions shown. The intersection of the isoclines denote possible equilibrium solutions for various combinations of null control and Nash control.

The stability of model (1), (2) in the neighborhood of an equilibrium solution is easily obtained from the linear equation describing motion in the vicinity of the equilibrium points. Let δx_1 and δx_2 be values of x_1 and x_2 in the neighborhood of an equilibrium point. From (1) and (2) with $H_1 = H_2 = 1$ it follows that

$$\overset{\circ}{\delta x}_1 = -x_1 \delta x_1 - x_1 \delta x_2 \tag{16}$$

$$\overset{\circ}{\delta x}_2 = \gamma\beta \left(\frac{x_2}{x_1}\right)^2 \delta x_1 - \gamma\beta \left(\frac{x_2}{x_1}\right)^2 \delta x_2 \tag{17}$$

where the coefficients of δx_1 and δx_2 are evaluated at the equilibrium point. The eigenvalues λ for this coefficient matrix are given by

$$2\lambda = -(x_1 + \gamma\beta x_2/x_1) \pm \sqrt{(x_1 + \gamma\beta x_2/x_1)^2 - 4\gamma\beta x_2(1 + x_2/x_1)} \tag{18}$$

It follows from (18) that all possible equilibrium points (in the triangle E, 5, 6, 7) are locally stable.

In particular, the Nash solution is locally stable. Assume then that the Nash solution represents the tentative limits on u_1 and u_2 chosen by the system manager. That is the prey are not endangered by a population size = .3536 and the predators are not endangered by a population size = .1464. The vulnerability of the system subject to the Nash limits on u_1 and u_2, i.e.,

$$0 < u_1 \leq .5 \tag{19}$$

$$0 \leq u_2 \leq .586 \tag{20}$$

will be obtained in the next section using the control theoretic methods previously mentioned.

Case 2. $H_1 = 1/N_1 = 1/kx_1$, $H_2 = 1/N_2 = a/r_1 x_2$ (Rate Harvesting)

At equilibrium, equations (1) and (2) reduce to

$$x_2 = 1 - x_1 - u_1/kx_1 \tag{21}$$

$$x_1 = \gamma x_2/(1 - au_2/r_1 x_2) \quad . \tag{22}$$

The intersection of these two isoclines represents equilibrium solutions. All possible positive equilibrium solutions are obtained by choosing $0 \leq u_1/k \leq 1/4$ and $0 \leq au_2/r_1 \leq 4\gamma + 2 - 4\gamma\sqrt{(\gamma+1)/\gamma}$. With $\gamma = 1$, all possible equilibrium solutions are contained in the triangle of Figure 2 composed of regions E, 5, 6, and 7.

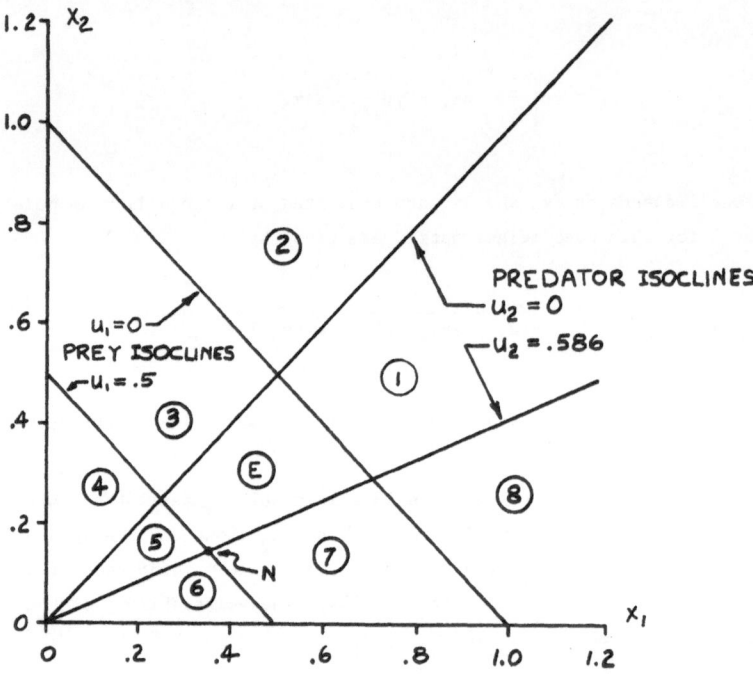

FIGURE 1.   Effort Harvest Isoclines for Null Control and Nash Control
N = Nash Equilibrium Point

Again, a tentative harvesting limit may be obtained by seeking a Nash solution for the yields (3) and (4). In this case, the yields are given by

$$y_1 = u_1/k = x_1(1 - x_1 - x_2) \tag{23}$$

$$y_2 = \beta au_2/r_1 = \beta x_2(1 - \gamma x_2/x_1) \quad . \tag{24}$$

A Nash solution for the controls u_1 and u_2 are subject to the constraints that an equilibrium point as given by (21) and (22) must exist. In this case, the Nash solution is not unique. For $\gamma = 1$, the set of Nash points illustrated in

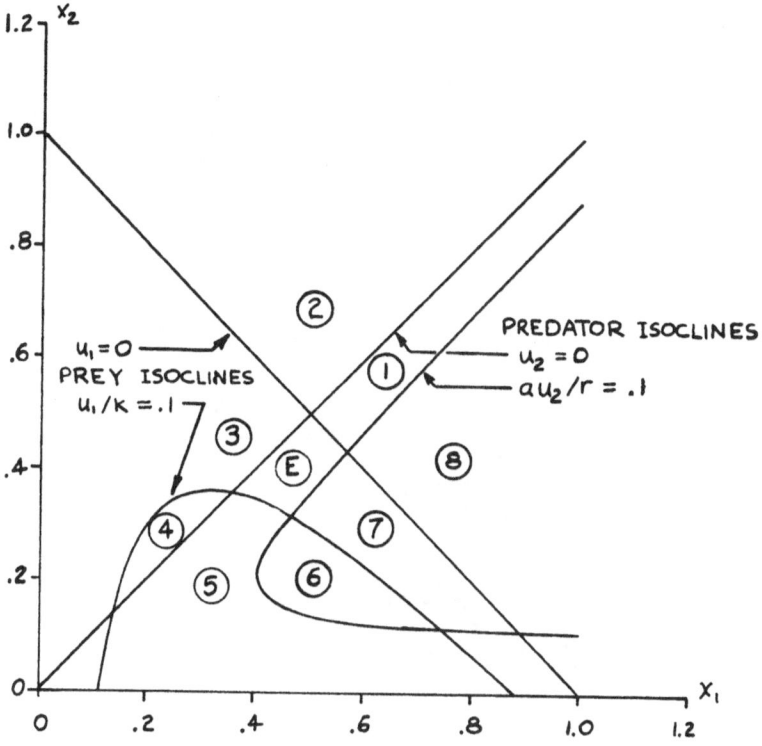

FIGURE 2. Rate Harvest Isoclines for Null Control and Constant Control
$u_1/k = au_2/r_1 = .1$

Figure 3 are obtained. Unfortunately, none of the Nash solutions illustrated in Figure 3 are suitable as a tentative harvesting limit. This follows from the fact that none of the Nash solutions are stable.

The linear equations describing motion in the vicinity of an equilibrium point for this case are given by

$$\overset{\circ}{\delta x}_1 = x_1 \frac{\partial f_1}{\partial x_1} \delta x_1 + x_1 \frac{\partial f_1}{\partial x_2} \delta x_2 \qquad (25)$$

$$\overset{\circ}{\delta x}_2 = x_2 \frac{\partial f_2}{\partial x_1} \delta x_1 + x_2 \frac{\partial f_2}{\partial x_2} \delta x_2 \qquad (26)$$

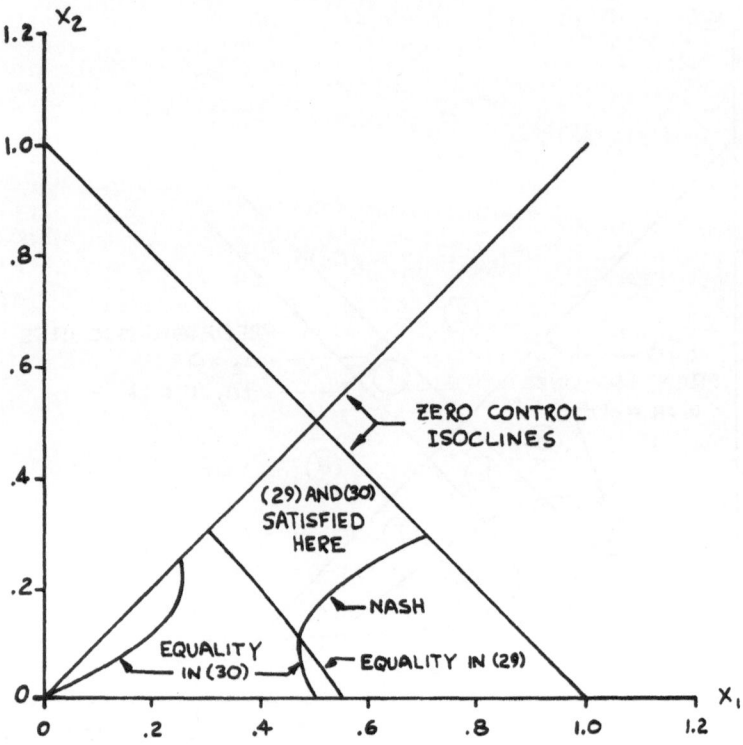

FIGURE 3. Stable Equilibrium Points for Rate Harvest

where δx_1 and δx_2 are values of x_1 and x_2 in the neighborhood of an equilibrium point,

$$f_1 = 1 - x_1 - x_2 - u_1/kx_1 \qquad (27)$$

$$f_2 = \beta(1 - \gamma x_2/x_1 - au_2/r_2 x_2) \ , \qquad (28)$$

and the coefficients of δx_1 and δx_2 are evaluated at the equilibrium point. Necessary and sufficient conditions for the eigenvalues of the coefficient matrix to have negative real parts are that

$$x_1 \frac{\partial f_1}{\partial x_1} + x_2 \frac{\partial f_2}{\partial x_2} < 0 \tag{29}$$

and

$$\frac{\partial f_1}{\partial x_1} \frac{\partial f_2}{\partial x_2} - \frac{\partial f_1}{\partial x_2} \frac{\partial f_2}{\partial x_1} > 0 \quad . \tag{30}$$

Since a necessary condition for a Nash solution is given by

$$\frac{\partial f_1}{\partial x_1} \frac{\partial f_2}{\partial x_2} = \frac{\partial f_1}{\partial x_2} \frac{\partial f_2}{\partial x_1} \tag{31}$$

it follows that the Nash solutions will be on the border between stable and unstable equilibrium points. Figure 3 illustrates those equilibrium points which satisfy both (29) and (30).

Since none of the Nash solutions are stable, some other equilibrium point must be chosen for setting a tentative harvesting limit. Consider for example tentative harvesting limits of $u_1/k = .1$ and $au_2/r = .1$. With $\gamma = 1$, the isoclines as defined by (21) and (22) under these limits intersect with the null harvesting isoclines as shown in Figure 2. The points of intersection are equilibrium points for the system (1), (2) under various combinations of constant control. There are two equilibrium points defined by the control

$$u_1/k = .1 \text{ and } au_2/r_2 = .1 \quad . \tag{32}$$

It follows from Figure 3 that the lower right hand equilibrium point is unstable. The other equilibrium point located at

$$x_1 = .465, \quad x_2 = .320 \tag{33}$$

is stable with the corresponding yields of

$$y_1 = .1, \quad y_2 = .1\beta \quad . \tag{34}$$

Assume then that this stable solution represents the one chosen by the system manager. Neither the prey nor predators are endangered by the population sizes given by (33). The vulnerability of the system subject to this choice, i.e.,

$$0 \leq u_1/k \leq .1 \tag{35}$$

$$0 \leq au_2/r_2 \leq .1 \tag{36}$$

may now be examined.

VULNERABILITY ANALYSIS

Case 1. $H_1 = H_2 = 1$ (Effort Harvesting)

Since all four equilibrium points defined by the isoclines of Figure 1 are stable, it seems reasonable that any of these equilibrium points could be reached by starting the system somewhere inside the trapezoid E defined by the isoclines. This is indeed the case. In fact, it is possible under appropriate manipulation of the control still satisfying (19) and (20) to drive the system outside the trapezoid. By employing conditions which must be satisfied on the boundary of all points reachable from the trapezoid a "worst case" harvesting program may be obtained. That is under a "worst case" program the system will be driven as far as possible from the trapezoid.

Suppose that the system were started on the boundary of the set of all points reachable from the trapezoidal region E and a control existed which would maintain the system on the boundary. Then this control must satisfy a Maximum principle (Grantham and Vincent, 1975). See also Grantham's paper in this volume.

In particular, there must exist multipliers λ_1 and λ_2 satisfying

$$\overset{\circ}{\lambda}_1 = -\partial H/\partial x_1, \qquad\qquad \overset{\circ}{\lambda}_2 = -\partial H/\partial x_2 \tag{37}$$

such that the control vector (u_1, u_2) maximizes

$$H = \lambda_1 x_1 (1 - x_1 - x_2 - u_1) + \lambda_2 \beta x_2 (1 - \gamma x_2/x_1 - u_2) \tag{38}$$

on the control set defined by (19) and (20) for every point on the boundary of the reachable set. Furthermore, the maximum value of H is zero.

By defining $P_1 = \lambda_1 x_1$, $P_2 = \beta \lambda_2 x_2$ the adjoint equations (37) reduces to

$$\overset{\circ}{P}_1 = P_1 x_1 - \gamma P_2 x_2/x_1 \tag{39}$$

$$\overset{\circ}{P}_2 = \beta(P_1 x_2 + \gamma P_2 x_2/x_1) \tag{40}$$

so that H is maximized by the controls

$$u_1 = \begin{cases} 0 \text{ if } P_1 > 0 \\ .5 \text{ if } P_1 < 0 \end{cases} \qquad u_2 = \begin{cases} 0 \text{ if } P_2 > 0 \\ .586 \text{ if } P_2 < 0 \end{cases} \tag{41}$$

with no possibility for singular control (i.e., $P_1 \equiv 0$ or $P_2 \equiv 0$). The structure of the control used on the boundary of the reachable set may be easily deduced by noting that when a prey isocline is crossed, P_2 must equal zero (since H = 0) and when a predator isocline is crossed, P_1 must equal zero. Thus u_1 switches when crossing a predator isocline and u_2 switches when crossing a prey isocline. By figuring out the control in any one region, the control for all other regions can be deduced from the above observation.

In the following discussion it is assumed that the regions defined in Figure 1 do not contain their boundary points. Suppose that the system is initially in E. Any other point in E can be reached by employing appropriate controls. To move outside this region a boundary must be crossed. Consider any point on the boundary of the region between E and 1. We note that

$$\overset{\circ}{x}_1 = \begin{cases} 0 \text{ if } u_1 = 0 \text{ (Definition of boundary } x_1 + x_2 = 1) \\ - \text{ if } u_1 = .5 \end{cases} \tag{42}$$

$$\overset{\circ}{x}_2 = \begin{cases} + \text{ if } u_2 = 0 \\ - \text{ if } u_2 = .586 \end{cases} \tag{43}$$

Thus to cross the boundary between E and 1 the control u_2 must be zero. The slope of the crossing trajectory is given by

$$\overset{\circ}{x}_2 / \overset{\circ}{x}_1 = - \frac{\beta}{u_1} \left(\frac{x_2}{x_1} \right) (1 - \gamma x_2 / x_1) \tag{44}$$

For $\gamma = 1$ and $.05 \le \beta \le .5$ (range of variables to be used) it is clear that this boundary can not be crossed with $u_1 = .5$. Since switching only takes place on iso-clines it is concluded that control for the boundary of the reachable set is $u_1 = u_2 = 0$ in region 1 and that the direction of motion is counterclockwise.

Consider now a trajectory following the boundary of the reachable set starting in region 1 with $u_1 = u_2 = 0$. Since $u_2 = 0$, the isocline separating regions 1 and

2 exists and the control u_1 switches to u_1 = .5 in region 2. Now since u_1 = .5 in region 2 there is no isocline separating regions 2 and 3 so that the control in region 3 is also given by u_1 = .5, u_2 = 0. Since u_1 = .5 in region 3, the isocline separating regions 3 and 4 exists and the control u_2 switches to u_2 = .586 in region 4. This process may be continued in order to map out the following table of control laws region by region.

TABLE 1.

Effort harvest controls for the boundary of the reachable set

Region	Control u_1	Control u_2
1	0	0
2	.5	0
3	.5	0
4	.5	.586
5	.5	.586
6	0	.586
7	0	.586
8	0	0

The boundary of the set of points reachable from the trapezoidal region E will be stable in the sense that if boundary control is used in the neighborhood of the boundary, then the system will asymptotically approach the boundary. By assuming that this local property is global a control law is easily formulated for throughout the state space. Namely use region 1 control throughout region 1, etc. It remains only to formulate a control law for region E. Any control law which will move the system out of E will do. For example, u_1 = u_2 = 0 will move the system out of E into 1. None of the other boundaries between E and the other regions can be crossed using this control.

Controls may now be assigned for the boundaries between E, 3; E, 5; and e, 7 which will guarantee the system to leave E. The following control algorithm for effort harvesting was obtained by this process.

$$F1Z = 1 - x_1 - x_2$$

$$F1M = F1Z - .5$$

$$F2Z = 1 - \gamma x_2/x_1$$

$$F2M = F2Z - .586$$

$$u_1 = \begin{cases} .5 \text{ if } F2M \leq 0 \text{ and } F1M > 0 \text{ or } F2Z \leq 0 \text{ and } F1M \leq 0 \\ 0 \text{ if above not true} \end{cases}$$

$$u_2 = \begin{cases} .586 \text{ if } F1M \geq 0 \text{ and } F2M < 0 \text{ or } F1Z \geq 0 \text{ and } F2M \geq 0 \\ 0 \text{ if above not true} \end{cases}$$

This control law is considered to be a "worst case" control law as it should (only necessary conditions were used) drive the system to the boundary of the set of points reachable from the region E.

Case 2. $H_1 = 1/N_1 - 1/kx_1$, $H_2 = 1/N_2 = a/r_1x_2$ (Rate Harvesting)

All four equilibrium points defining the corners of region E in Figure 2 are stable. As in the previous case if the system starts somewhere inside of E, it is possible under appropriate manipulation of the control satisfying (35) and (36) to drive the system outside of E. The boundary of all points reachable from E are again obtained using the Maximum principle. In this case

$$H = \lambda_1 x_1(1 - x_1 - x_2 - u_1/kx_1) + \lambda_2 \beta x_2(1 - \gamma x_2/x_1 - au_2/r_1x_2) \quad . \tag{45}$$

Applying the maximum principle, it is again found that u_1 switches when crossing a predator isocline and u_2 switches when crossing a prey isocline. By using the same arguments as in the previous case, the control to be used on the boundary of the reachable set when located in the various regions is obtained as summarized in Table 2.

TABLE 2.   Rate Harvest Controls for the boundary of the reachable set

REGION	CONTROL u_1/k	CONTROL au_2/r_1
1	0	0
2	.1	0
3	.1	0
4	.1	.1
5	.1	.1
6	0	.1
7	0	.1
8	0	0

By applying region 1 control throughout region 1, etc. and formulating a control law for E as before, the following "worst case" control algorithm for rate harvesting was obtained

$$F1Z = 1 - x_1 - x_2$$

$$F1M = F1Z - .1/x_1$$

$$F2Z = 1 - \gamma x_2/x_1$$

$$F2M = F2Z - .1/x_2$$

$$u_1 = \begin{cases} .1 \text{ if } F2M \leq 0 \text{ and } F1M > 0 \text{ or } F2Z \leq 0 \text{ and } F1M \leq 0 \\ 0 \text{ if above not true} \end{cases}$$

$$u_2 = \begin{cases} .1 \text{ if } F1M \geq 0 \text{ and } F2M < 0 \text{ or } F1Z \geq 0 \text{ and } F2M \geq 0 \\ 0 \text{ if above not true} \end{cases}$$

RESULTS

Case 1. Effort Harvesting

The result of employing the worst case effort harvesting program is illustrated in Figures 4, 5, and 6 for Nash limits (19) and (20) on the control, $\gamma = 1$ and $\beta = .05, .1$ and $.5$ respectively. For this case, whether one starts inside or outside the boundary shown, the worst case control will ultimately move the system to the boundary. The fact that one can start anywhere in the positive orthant outside the boundary and under worst case harvesting still reach the boundary illustrates considerable inherent stability of equations (1) and (2) under effort harvesting. None of the species can be driven to extinction using Nash limits on the controls.

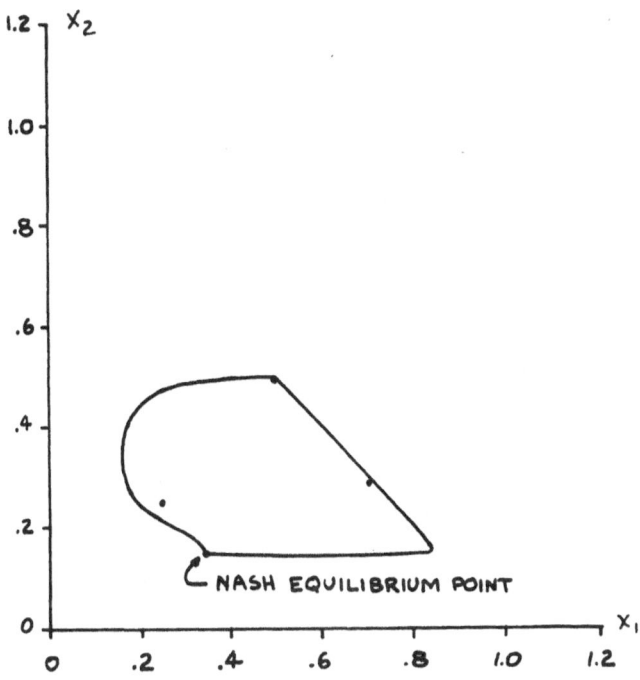

FIGURE 4.   Effort Harvest Vulnerability under Nash Limits, $\gamma = 1$, and $\beta = 0.05$

It was assumed that none of the species were considered to be endangered at the Nash equilibrium point. Note that in each case the prey could be driven to population levels lower than the Nash level, possibly to an endangered level. In other words, if the Nash equilibrium solution was just marginally above an endangered criteria, then this solution for harvesting limits should be discarded by the manager of this renewable resource ecosystem for one in which the system can not dynamically violate the endangered species requirement.

It is of interest to note that the boundary of the reachable set is relatively

insensitive to β. This result could have considerable bearing on accuracy requirements for the data.

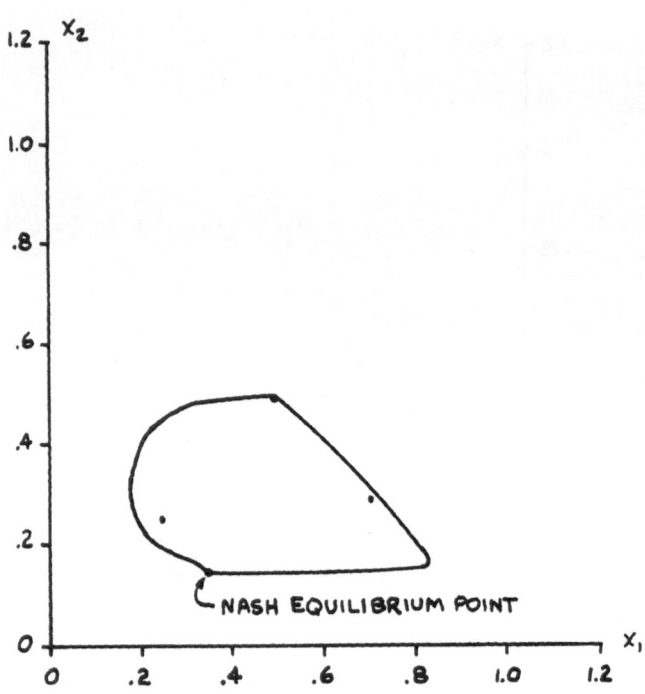

FIGURE 5.   Effort Harvest Vulnerability under Nash Limits, $\gamma = 1$, and
$\beta = 0.10$

Note also that the equilibrium points may or may not form a part of the reachable set boundary. For example, the Nash equilibrium point is on the boundary for
$\beta = .05$ and $\beta = .1$, but is interior to the set for $\beta = .5$. This result is related
to whether the corresponding eigenvalues are real or not. Eigenvalues for the four
equilibrium points may be obtained from (18) as summarized Table 3.

Note that the eigenvalues at the Nash equilibrium point are real for $\beta = .05$,
$\beta = .1$, and complex for $\beta = .5$. Examining the eigenvalues at the other equilibrium
points and noting their location relative to the boundary in Figures 4, 5, and 6
it is found that an equilibrium point will be inside the boundary of the reachable
set if the eigenvalue is complex and may or may not be on the boundary of the

reachable set if the eigenvalue is real. (In all cases the eigenvalues have negative real parts.)

TABLE 3

Equilibrium point eigenvalues under effort harvesting

CONTROL		EQUILIBRIUM POINT		EIGENVALUES		
u_1	u_2	x_1	x_2	$\beta = .05$	$\beta = .1$	$\beta = .5$
0	0	.500	.500	real	complex	complex
0	.586	.707	.293	real	real	real
.5	0	.250	.250	complex	complex	complex
.5	.586	.354	.146	real	real	complex

Case 2. Rate Harvesting

Figure 7 illustrates the results of employing worst case rate harvesting with
control limits $u_1/k = au_2/r_2 = .1$, $\gamma = 1$ and $\beta = .1$. The inner region A contains
the four stable equilibrium points. Region A has the property that under worst
case rate harvesting, if the system starts in A, then it will ultimately be driven
to the boundary of A. The next region B has the property that under worst case
harvesting, if the system starts in B, then it will be ultimately driven to the
boundary of A. Region C has the property that under worst case harvesting, if the
system starts in C, then one or the other of the species will be driven to extinction. Thus unlike the effort harvesting case, under rate harvesting the boundary
of region A is stable only in a limited neighborhood of the boundary.

Note that the boundary between regions A and B represents the boundary of the
set of points reachable from within A. Whereas the boundary between B and C represents the boundary of the set of points controllable to the set A. In each case
system trajectories which lie on these boundaries must satisfy a maximum principle.
The worst case rate harvesting algorithm results from the necessary conditions for
either case. Hence the same algorithm was used to generate both boundaries.

Note that the lower right hand unstable equilibrium point corresponding to constant control $u_1/k = au_2/r_2 = .1$ lies on the boundary between regions B and C.

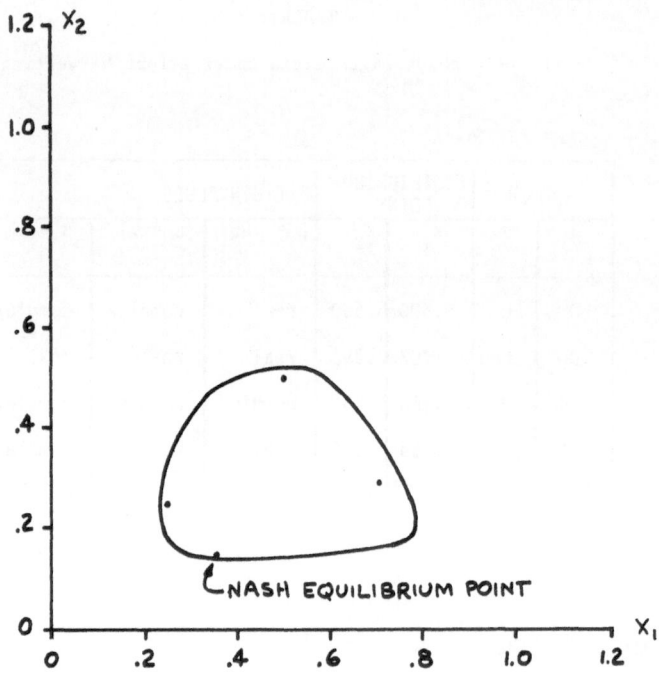

FIGURE 6.   Effort Harvest Vulnerability under Nash Limits, $\gamma = 1$, and
$\beta = .5$

It follows from Figure 7 that both the prey and predators can be driven to
values less than those corresponding to the constant control $(u_1/k = au_2/r_2 = .1)$
equilibrium point, possibly endangering one or both of the species. It is of
interest to note that under worst case rate harvesting, it is possible to drive
one of the species to extinction even when the system starts from a seemingly
favorable abundance of predators (from the harvesters' point of view) such as
point S in region C.

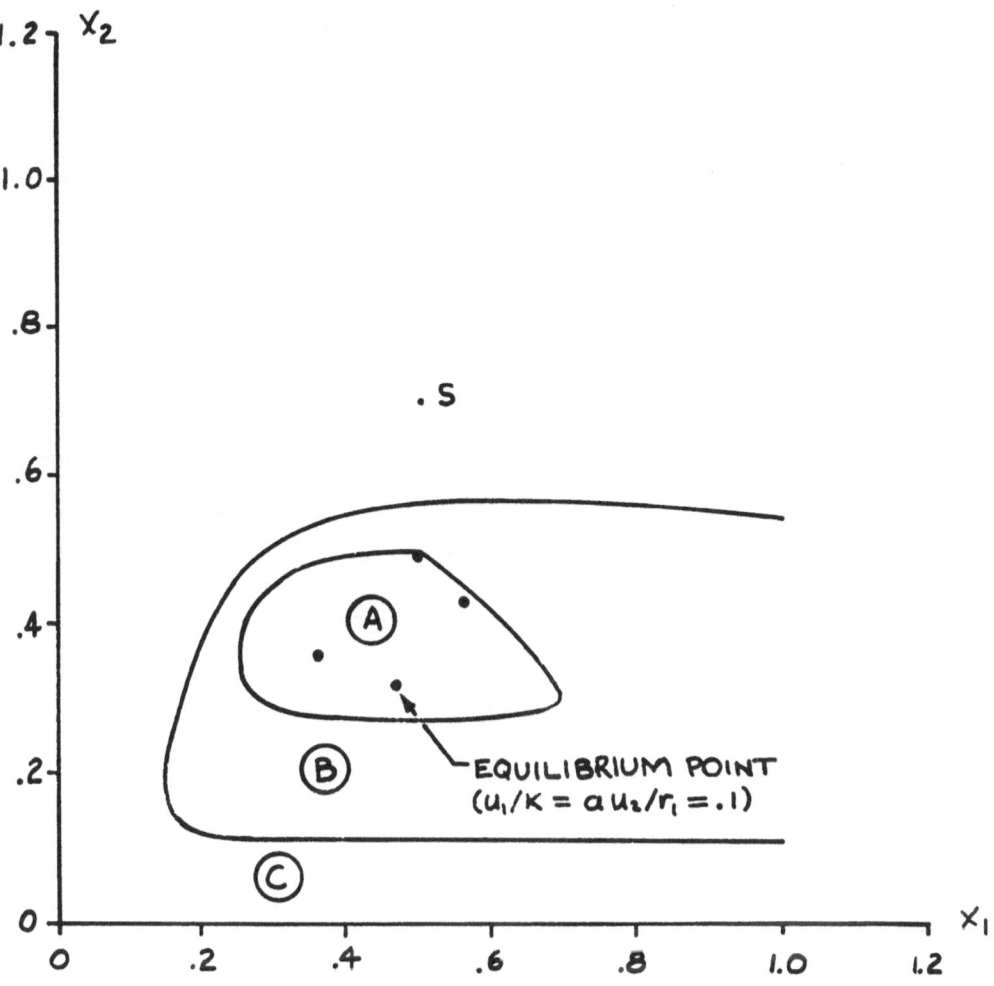

FIGURE 7.   Rate Harvest Vulnerability under Limits
$0 \leq u_1/k \leq .1$ and $0 \leq au_2/r_1 \leq .1$

REFERENCES

May, R.M., Beddington, J.R., Clark, C.W., Holt, S.J., and Laws, R.M. 1979. "Management of Multispecies Fisheries," Science, 205, pp. 267-277

Beddington, J.R. and Cooke, J.G. 1979. "Harvesting From a Predator-Prey System," Submittted to Mathematical Biosciences

Beddington, J.R. and May, R.M. 1977. "Harvesting Natural Populations in the Randomly Fluctuating Environment," Science, 197, pp. 463-465

Goh, B.S., Leitmann, G., and Vincent, T.L. 1974. "Optimal Control of a Prey-Predator System," Mathematical Biosciences, 19, pp. 263-286

Vincent, T.L. 1975. "Pest Management via Optimal Control Theory," Biometrics, 31, pp. 1-10

Vincent, T.L., Cliff, E.M., and Goh, B.S. 1974. "Optimal Direct Control Programs for a Prey-Predator System," J. Dynamical Systems Measurement and Control," March, pp. 71-76

Vincent, T.L. and Grantham, W.G. 1980. Optimality in Parametric Systems, Wiley, In Press

Goh, B.S. 1976. "Nonvulnerability of Ecosystems in Unpredictable Environments," Theoretical Population Biology, 10, pp. 83-95

Goh, B.S. 1977. "Stability in a Stock-Recruitment Model of an Exploited Fishery," Mathematical Biosciences, 33, pp. 359-372

LaSalle, J. and Lefschetz, S. 1961. Stability by Liapunov's Direct Method, Academic Press, New York

Vincent, T.L. and Anderson, L.R. 1979. "Return Time and Vulnerability For a Food Chain Model," Theoretical Population Biology, 15, pp. 217-231

Grantham, W.J. and Vincent, T.L. 1975. "A Controllability Minimum Principle," J. Optimization Theory and Applications, 17, pp. 93-114

NONVULNERABILITY OF TWO SPECIES INTERACTIONS

M. E. Fisher and B. S. Goh
Mathematics Department
University of Western Australia
Nedlands W.A. 6009
Australia

The standard stability analysis of an ecosystem model assumes that the model is not disturbed after an initial perturbation. But ecosystems in the real world are subjected to frequent disturbances. The concept of nonvulnerability was developed to describe the ability of a dynamical system to remain healthy under continual disturbances.

We shall discuss some applications of optimal control theory for studying the nonvulnerability of two species interactions which are subjected to continual and unpredictable, but bounded disturbances.

INTRODUCTION

Standard stability analyses of ecosystem models, whether they be of a local or global nature, assume that the model is not disturbed after an initial perturbation. Practical ecosystems are, however, continually disturbed by unpredictable forces due to changes in climatic conditions, migrating species, diseases etc. The concept of nonvulnerability was developed (Goh, 1975, 1976) to describe the ability of an ecosystem model to remain healthy under continual and unpredictable disturbances. Goh (1979) applied Liapunov function methods to carry out a conservative analysis of Lotka-Volterra models and this approach is also applicable to a wider class of nonlinear models. Goh (1976) also suggested that optimal control theory could be used to study the nonvulnerability of systems.

In this paper, we shall apply some of the techniques of optimal control theory to the study of nonvulnerability of two species interactions. We utilize the relationship between nonvulnerability and the control concepts of reachability and controllability (Grantham and Vincent, 1975; Vincent and Anderson, 1979). Problems associated with computing reachable sets for nonlinear ecosystem models will be discussed and illustrated using specific competition, prey-predator and host-parasite models.

Finally, we look at ecosystem models in which one or both species is subjected to optimal harvesting. Some examples of the effects of continual disturbances on these models are presented.

DEFINITIONS AND CONCEPTS

Consider the ecosystem model

$$\dot{N}_i = G_i(N_1,N_2\ldots,N_m) \ , \ i = 1,2,\ldots,m \ , \tag{1}$$

which is assumed to have a unique nontrivial equilibrium $N^* = (N_1^*,N_2^*,\ldots,N_m^*)$. We are interested in the effects of continual, but unpredictable, disturbances $u(t) = (u_1(t),u_2(t),\ldots,u_n(t))$ on the system (1). It will be assumed that these disturbances are piecewise continuous functions and that it is possible to make a priori estimates on their bounds, i.e., there exists constants m_r and M_r such that

$$m_r \leq u_r(t) \leq M_r \ , \ r = 1,2,\ldots,n \ . \tag{2}$$

In what follows, we will refer to the set of admissable disturbance functions satisfying (2) as the set U and the model equations in their most general form are given by

$$\dot{N}_i = F_i(N_1,N_2,\ldots,N_m \ ; \ u_1,u_2,\ldots,u_n) \ , \ i = 1,2,\ldots,m \ . \tag{3}$$

Our definition of nonvulnerability follows that of Goh (1976).

Definition: Given a class of disturbance functions U , $t_1 > 0$, and sets $S \subset \mathbf{R}^m$ and $Z \subset \mathbf{R}^m$, the system (3) is said to be nonvulnerable with respect to $\{U,S,Z,t_1\}$ if, for any trajectory $N(t)$, the condition $N(0) \ \epsilon \ S$ implies that $N(t) \notin Z$ for any $t \ \epsilon \ [0,t_1]$ and $u(t) \ \epsilon \ U$, i.e., there is no $u(t) \ \epsilon \ U$ which can drive the system from an initial state in S to a state in Z in the time interval $[0,t_1]$. Other related types of stability which have been referred to in the literature are: practical stability (LaSalle and Lefshetz, 1961), stability under perturbing forces (Weiss and Infante, 1965) and total stability (Hahn, 1967).

Liapunov function methods have previously been used to obtain sufficient conditions for nonvulnerability. These methods are essentially based on the following result:

Let $V(N)$ be a scalar function satisfying the following

(a)   V is positive definite with $V(N^*) = 0$;

(b)   there exists a number $p > 0$ such that, for all

$N \ \epsilon \ P = \{N | V(N) = p\}$ and all $u(t) \ \epsilon \ U$, $\dot{V} = \nabla V \cdot F(N,u) \leq 0$.

Then if $S \subset P \subset Z^c$, the system (3) is nonvulnerable with respect to $\{U,S,Z,t_1\}$ for any $t_1 > 0$.

Sufficient conditions based on this and similar results have been obtained by Goh (1976) for the nonvulnerability of Lotka-Volterra models and by Harrison (1979) for the persistence of these models under certain classes of disturbances.

The concept of nonvulnerability is closely related to the ideas of reachability and controllability in control theory.

<u>Definition</u>: Given a class of control functions U, $t_1 > 0$, and an initial set S, we define the set $R(U,S,t_1)$ to be the totality of points which can be reached from the set S under all possible control functions u(t) ϵ U for t ϵ $[0,t_1]$.

<u>Definition</u>: Given a class of control functions U , $t_1 > 0$, and a terminal set Z, the set $C(U,Z,t_1)$ is defined to be the set of all points N(0) for which there exists a control u(t) ϵ U such that N(t) ϵ Z for some t ϵ $[0,t_1]$.

An immediate consequence of these definitions is that, for $t_1 > 0$ and given sets U,S and Z, we can now determine whether or not system (3) is nonvulnerable with respect to $\{U,S,Z,t_1\}$ by computing either $R(U,S,t_1)$ or $C(U,Z,t_1)$.

One can see that the Liapunov function approach gives us an estimate of the set $R(U,S,t_1)$. This estimate is usually a conservative one and of course depends on the particular Liapunov function employed. In the remainder of this paper, we shall focus our attention on the problem of computing the sets $R(U,S,t_1)$ and $C(U,Z,t_1)$ for two species interactions of the form

$$\dot{N}_i(t) = f_i(N_1,N_2) + u_i(t)N_i , \quad i = 1,2 . \tag{4}$$

Equation (4) is a special case of equation (3) in which we have subjected the ecosystem model to density independent disturbances acting on the growth rate of each species.

FINITE TIME NONVULNERABILITY

Suppose that the sets S and Z can be defined analytically by some suitable function h : $\mathbb{R}^2 \to \mathbb{R}$, i.e., there exists constant h_S and h_Z such that

$$S = \{N | h(N) \le h_S\} ,$$
$$Z = \{N | h(N) \ge h_Z\} .$$

For an ecosystem model, S is some neighbourhood of the undisturbed equilibrium N^* containing, in some sense, desirable or healthy states of the system and Z is a set of undesirable or unhealthy states. A convenient function for describing S and Z is given by

$$h(N) = \sum_{i=1}^{2} [N_i/N_i^* - 1 - \ln(N_i/N_i^*)] \quad . \tag{5}$$

h has a global minimum of zero at N^*, and is a "measure" of the distance of the two populations from their equilibrium values.

The Reachable Set $R(U,S,t_1)$

For a given $t_1 > 0$ we will now compute $R(U,S,t_1)$, the totality of points which can be reached from S in the time interval $[0,t_1]$ under all possible $u(t) \in U$. To this end we consider the following control problem:

System: $\dot{N}_i = f_i(N_1,N_2) + u_i(t)N_i, \quad i = 1,2$;

Initially: $h(N(0)) = h_S$;

Terminally: $N(T)$ is "given" ;

Constraints: $m_i \leq u_i(t) \leq M_i, \quad i = 1,2$;

Objective: minimise T .

The Hamiltonian function for this problem is defined as

$$H(N,u,p) = \sum_{i=1}^{2} p_i(t)[f_i(N_1,N_2) + u_i(t)N_i] \quad . \tag{6}$$

The $p_i(t)$ are costate variables satisfying the equations

$$\dot{p}_i = - \frac{\partial H}{\partial N_i}$$

$$= - \sum_{j=1}^{2} p_j(t) \left[\frac{\partial f_j}{\partial N_i} + u_j(t)\delta_{ij} \right], \quad i = 1,2 , \tag{7}$$

where δ_{ij} is the kronecker delta function. A minimum-time control $u^*(t)$ is determined by minimising H with respect to u subject to the constraints. This leads to

$$u_i^*(t) = \begin{cases} m_i \text{ only if } p_i(t) > 0 ; \\ \text{singular control only if } p_i(t) = 0 ; \\ M_i \text{ only if } p_i(t) < 0 . \end{cases} \tag{8}$$

Finally, the transversality conditions imply that

$$p_i(0) = - \frac{\partial h}{\partial N_i} (N(0)) \bigg/ \sum_{j=1}^{2} \frac{\partial h}{\partial N_j} (N(0))[f_j(N(0)) + u_j^*(0)N_j(0)], \quad i = 1,2 . \tag{9}$$

For this to be a well defined optimal control problem we impose the additional condition that for all initial points $N(0)$, there exists a control function $u(t) \in U$ such that

$$\nabla h \cdot N > 0 \quad . \tag{10}$$

This guarantees that S is a subset of $R(U,S,t_1)$ and that all minimum time trajectories leave S and cannot re-enter S at a later time.

We now compute the minimum time trajectories for initial values on $h(N) = h_S$ be integrating equations (4) and (7), using condition (8), subject to the initial (9). Consider all the points which can be reached in the same minimum time t_1. These points form a closed curve and correspond to the boundary of the set $R(U,S,t_1)$.

Example 1. Consider the prey-predator model

$$\dot{N}_1 = N_1[2 + N_1(4 - N_1)/(1 + 2N_1) - N_2/(1 + 0\cdot05N_1)] + u_1N_1 \, ,$$
$$\tag{11}$$
$$\dot{N}_2 = N_2[-0\cdot4156 + N_2(1 - N_2)/(1 + 8N_2) + 0\cdot1N_1/(1 + 0\cdot05N_1)] + u_2N_2 \quad .$$

The prey population, N_1, in this model sustains an Allee effect and the per capita death rate of the predator is a nonlinear function of its density. The undisturbed equilibrium $N^* = (6,1.4)$ is globally stable (Goh, 1978). Let the boundary of S be described by the curve $h(N) = h_S$ which passes through the points $(0\cdot95N_1^*,N_2^*)$ and $(N_1^*,0\cdot95N_2^*)$, i.e., $S = \{N|h(N) \le 0\cdot00129\}$. Figure 1 shows some minimum time trajectories initiating from the boundary of S for the case $-0\cdot3 \le u_i(t) \le 0\cdot3$, $i = 1,2$. The reachable sets are illustrated for $t_1 = 1\cdot0, 2\cdot0$ and $4\cdot0$. For this example $R(U,S,t_1)$ will eventually intersect the $N_2 = 0$ axis for t_1 large enough.

The Controllable Set $C(U,Z,t_1)$

Instead of computing the reachable set $R(U,S,t_1)$ to determine whether or not a system is nonvulnerable, we could compute the controllable set $C(U,Z,t_1)$. For given sets U,S,Z and $t_1 > 0$ the system (4) is nonvulnerable with respect to $\{U,S,Z,t_1\}$ if the intersection of S with $C(U,Z,t_1)$ is empty. The sets $C(U,Z,t_1)$ are obtained by solving the same minimum time control problem with different end conditions.

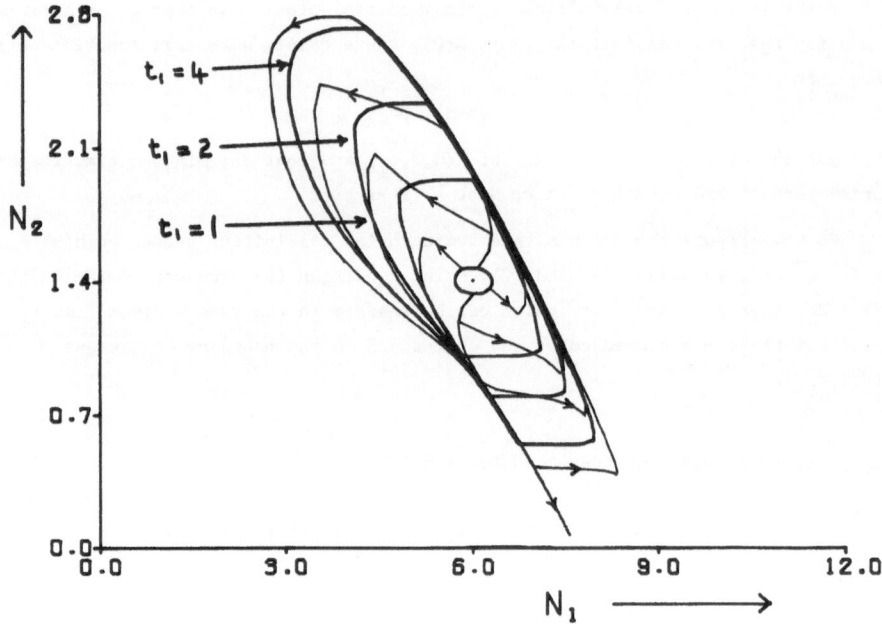

FIGURE 1. Reachable sets $R(U,S,t_1)$ for example 1

System:	$\dot{N}_i = f_i(N_1,N_2) + u_i(t)N_i$, $i = 1,2$;
Initially:	$N(0)$ is "given" ;
Terminally:	$h(N(T)) = h_Z$;
Constraints:	$m_i \le u_i(t) \le M_i$, $i = 1,2$;
Objective:	minimise T .

The Hamiltonian function, costate equations and optimality conditions are the same as previously. The terminal conditions on the costate variables are

$$P_i(T) = -\frac{\partial h}{\partial N_i}(N(T)) \bigg/ \sum_{j=1}^{2} \frac{\partial h}{\partial N_j}(N(T))[f_j(N(T)) + u_j^*(T)N_j(T)], \quad i = 1,2 . \qquad (12)$$

To find the minimum time trajectories which terminate on $h(N(T)) = h_Z$ we integrate the system and costate equations backwards in time from the terminal surface using the optimality condition (8) to determine $u^*(t)$. Since we are only interested in finding that part of $C(U,Z,t_1)$ which lies outside Z, we need only consider that part of the surface $h(N) = h_Z$ for which there exists a control function $u(t) \varepsilon \, U$ such that

$$\nabla h \cdot \dot{N} > 0 \tag{13}$$

One advantage of this approach over that of using reachable sets is that we need only compute trajectories for which (13) is satisfied on the terminal surface. In practice only a small part, if any, of the surface $h(N) = h_Z$ will have this property. Hence the amount of computation involved in computing $C(U,Z,t_1)$ may be significantly less than in computing $R(U,S,t_1)$. Of course if there is no point on $h(N) = h_Z$ satisfying (13) then $C(U,Z,t_1) = Z$.

Figure 2 shows the controllable set $C(U,Z,t_1)$ for $t_1 = 5$ and some minimum time trajectories for the prey-predator model of example 1 with $-0 \cdot 3 \leq u_i(t) \leq 0 \cdot 3$ for $i = 1,2$. Here the set Z is defined by $\{N | h(N) \geq h_Z\}$ with $h_Z = 0 \cdot 809$. The boundary of Z passes through the points $(0 \cdot 2N_1^*, N_2^*)$ and $(N_1^*, 0 \cdot 2N_2^*)$.

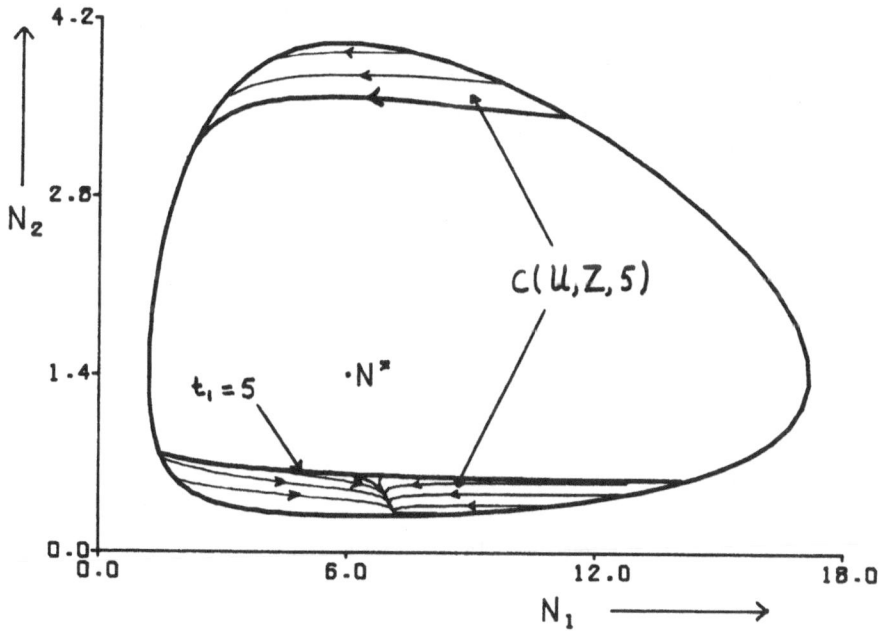

FIGURE 2. The controllable set $C(U,Z,5)$ for example 1

INFINITE TIME NONVULNERABILITY

Given a class of disturbance functions U and sets S and Z, the system (4) is said to be nonvulnerable with respect to $\{U,S,Z\}$ if there is no $u(t) \in U$ which can drive the system from an initial state in S to a state in Z for any $t > 0$. Similarly, we refer to $R(U,S)$ as the totality of points which can be reached from S under any $u(t) \in U$ and $t > 0$, and $C(U,Z)$ as all those points which can be driven to Z for any $u(t) \in U$ and $t > 0$.

If we have condition (10) satisfied on the boundary of S then $R(U,S)$ will be the same as the reachable set from the equilibrium point N^*, i.e., $R(U,S) = R(U,N^*)$. One method of determining this reachable set, as suggested by Vincent and Anderson (1979), is to compute minimum time trajectories from the equilibrium point following the procedure outlined in the previous section. The only difference is that the initial values of p_1 and p_2 must be consistent with the transversality condition,

$$p_1(0)u_1^*(0)N_1^* + p_2(0)u_2^*(0)N_2^* = -1 , \tag{14}$$

with $u_1^*(0)$ and $u_2^*(0)$ given by equation (8). Curves of constant minimum time form closed curves around the equilibrium point and the boundary of $R(U,N^*)$ corresponds to the limiting curve as the minimum time tends to infinity. There appears to be computational difficulties associated with this approach, and we will discuss some of these shortly.

If condition (10) is satisfied on only part of the boundary of S, then we have either of two possible situations arising. Either S is contained in $R(U,N^*)$, in which case $R(U,S) = R(U,N^*)$, or part of S lies outside $R(U,N^*)$. In the second case there may exist minimum time trajectories leaving S which will trace out the boundary of $R(U,S)$. We will restrict our attention to when $R(U,S) = R(U,N^*)$ and the reader is referred to Grantham and Vincent (1975) for discussion on a problem related to the other case.

The reachable set $R(U, N^*)$ is of course independent of any performance criteria. There appears to be no reason therefore why we could not replace T in the optimal control problem by any reasonably well behaved objective function. In fact numerical experiments were conducted in which we minimised

$$\int_0^T [u_1^2(t) + u_2^2(t)] \, dt .$$

The optimal control function for this problem consists of a mixture of interior and boundary controls. The resulting trajectories exhibit similar properties to those of the minimum time problem; and from them, one is able to obtain an estimate of $R(\mathcal{U},N^*)$.

We now return to the problem of computing trajectories for the minimum time control problem and the determination of the set $R(\mathcal{U},N^*)$. If constant control is applied in which $u_1(t)$ and $u_2(t)$ are set to either their minimum or maximum values then, provided the control is small enough, the system will approach a new equilibrium which we shall call a "disturbed" equilibrium point. For some ecosystem models these four disturbed equilibrium points lie on the boundary of $R(\mathcal{U},N^*)$ and for some models they lie in the interior. A particular minimum time trajectory leaving N^* has the property that it actually traces out the boundary of $R(\mathcal{U},N^*)$, as it asymptotically approaches it, or it approaches one of the disturbed equilibrium points lying on the boundary of $R(\mathcal{U},N^*)$. In the second case, there is no single trajectory which will map out the boundary of the reachable set.

On the basis of numerical experiments with a variety of competition, mutualistic, prey-predator, and host-parasite models we will attempt to categorise the ecosystem models given by equation (4) into one of three classes. These classes are described in terms of the properties of the disturbed equilibrium points and their relationship to the boundary of the reachable set.

Class 1. This class of models forms a subset of all prey-predator and host-parasite models and has the property that the four disturbed equilibrium points all have complex eigenvalues. These disturbed equilibria all lie in the interior of $R(\mathcal{U},N^*)$, and to determine the reachable set we need only a single minimum time trajectory which will trace out the boundary of $R(\mathcal{U},N^*)$ as it asymptotically approaches it. The following is an example of an ecosystem model in this class and is illustrated in Figure 3.

Example 2.

$$\dot{N}_1 = N_1(2 - 0\cdot02N_1 - 0\cdot1N_2/(1 + 0\cdot02N_1)) + u_1N_1 ,$$

$$\dot{N}_2 = N_2(-0\cdot5 + 0\cdot02N_1/(1 + 0\cdot02N_1)) + u_2N_2 ,$$

$$-0\cdot1 \le u_i(t) \le 0\cdot1 , \quad i = 1,2 .$$

In this prey-predator model, the predator has a type II functional response, and the undisturbed equilibrium is at $N^* = (50,20)$. The four disturbed equilibrium points are $(33\cdot3,23\cdot9)$, $(75,15)$, $(33\cdot3,20\cdot6)$, $(75,10)$.

Class 2. For these models the disturbed equilibria have real eigenvalues and they all lie on the boundary $R(U,N^*)$. This class includes all competition and mutual- istic models. The minimum time trajectories from N^* converge to one of two of the disturbed equilibria in the limit as the minimum time tends to infinity. A prop- erty of these models is that it can be computationally difficult to generate mini- mum time trajectories which reach some parts of $R(U,N^*)$, even using double pre- cision arithmetic. There are initial values for the costate variables $p_1(t)$ and $p_2(t)$ close to which a small change in the initial conditions can make a large

change in the resultant minimum time trajectory. A significant proportion of $R(U,N^*)$ can only be reached by trajectories with initial values of $p_1(t)$ and $p_2(t)$ very close to these values. There is, however, a simpler and far cheaper means of generating the boundary of $R(U,N^*)$ for this class of models. This is by integrating the system of equations (4) under constant control starting from the disturbed equilibrium points. The boundary of the reachable set $R(U,N^*)$ will then be traced out by two constant control trajectories initiating from each of two disturbed equilibrium points.

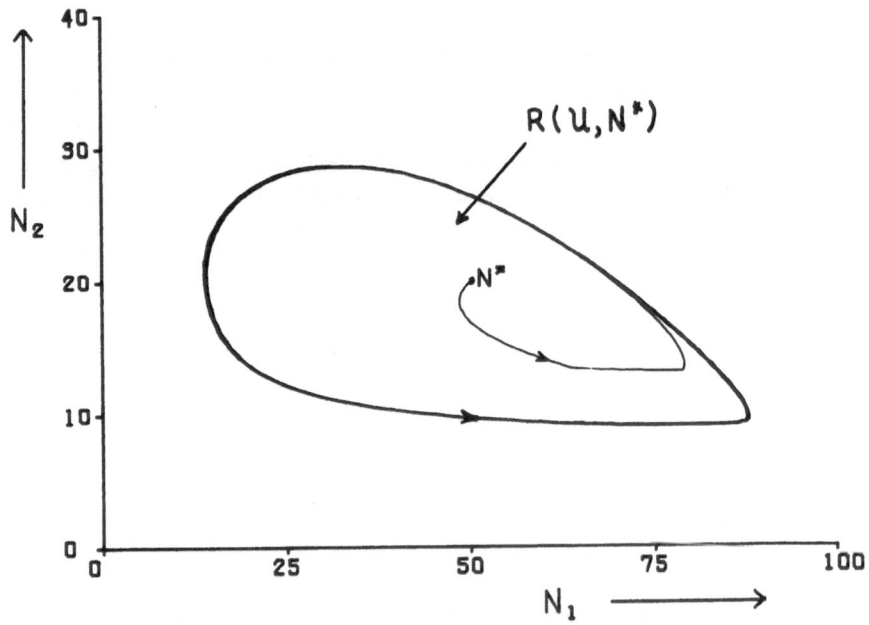

FIGURE 3. The reachable set $R(U,N^*)$ for example 2

Example 3.

$$\dot{N}_1 = N_1[1\cdot5 - 1\cdot5N_1^{0\cdot35} - 0\cdot6N_2] + u_1N_1 .$$

$$\dot{N}_2 = N_2[4 - 0\cdot6N_1 - 4N_2^{0\cdot12}] + u_2N_2 ,$$

$$- 0\cdot1 \leq u_i(t) \leq 0\cdot1 ; \quad i = 1,2 .$$

This is the competition model of Gilpin and Ayala (1973). The parameter values have been chosen to correspond closely to those provided by Gilpin and Ayala for two competing Drosophila populations. The species densities have been scaled with respect to their carrying capacities. The undisturbed equilibrium, N^*, is at the point $(0\cdot5249, 0\cdot5049)$, and the disturbed equilibrium points are $(0\cdot6106, 0\cdot5632)$, $(0\cdot9398, 0\cdot2204)$, $(0\cdot1654, 1\cdot0016)$, $(0\cdot4471, 0\cdot4472)$. The boundary of $R(U,N^*)$ is generated by constant control trajectories initiating from the points $(0\cdot9398, 0\cdot2204)$ and $(0\cdot1654, 1\cdot0016)$. These, together with some minimum time trajectories, are illustrated in Figure 4.

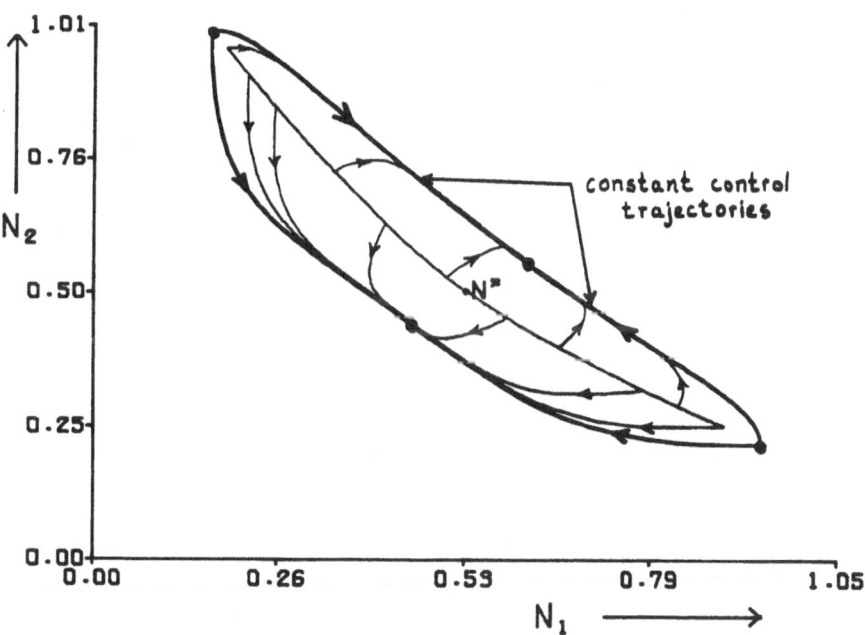

FIGURE 4. The reachable set $R(U,N^*)$ for example 3

<u>Class 3</u>. For this class of models, the disturbed equilibria will have real eigen-values or a mixture of real and complex eigenvalues associated with them. The property that distinguishes them from the class 2 models is that some of the dis-turbed equilibria lie on the boundary of $R(U,N^*)$ and the others are in the interi-or. On the basis of experiment, this class appears to consist of those prey-pred-ator models not in class 1.

As for competition models, it can be computationally difficult to obtain min-imum time trajectories which reach certain parts of $R(U,N^*)$ due to the sensitivity of the trajectories to the initial conditions. All the minimum time trajectories ultimately approach the disturbed equilibria lying on the boundary of the reachable set. These disturbed equilibria must have real eigenvalues since those with com-plex eigenvalues must lie in the interior of the reachable set, although they may be very close to the boundary (see example 4). The simple means of generating the boundary of $R(U,N^*)$ for class 2 models fails for this class of models. This is because there are some parts of the boundary which cannot be traced out by constant control trajectories initiating from the disturbed equilibrium points.

Figure 5(a) shows the reachable set $R(U,N^*)$ and some minimum time trajectories for the prey-predator model of example 1 with $-0 \cdot 15 \leq u_i(t) < 0 \cdot 15$, for $i = 1,2$. The disturbed equilibria, which all have real eigenvalues, are located at the points $(4 \cdot 982, 2 \cdot 128)$, $(7 \cdot 401, 0 \cdot 763)$, $(4 \cdot 619, 1 \cdot 934)$, $(7 \cdot 057, 0 \cdot 572)$. Also shown in Figure 5(a) are some constant control trajectories initiating from the disturbed equilibrium points.

<u>Example 4</u>.

$$\dot{N}_1 = N_1(2 - N_1 - N_2) + u_1 N_1 ,$$

$$\dot{N}_2 = N_2(-0 \cdot 1 + 0 \cdot 1 N_1) + u_2 N_2 ,$$

with $\quad - 0 \cdot 05 \leq u_i(t) \leq 0 \cdot 05 \quad$ for $\quad i = 1,2$.

This is a Volterra prey-predator model with $N^* = (1,1)$. The disturbed equilibria are located at the points $(0 \cdot 5, 1 \cdot 55)$ and $(0 \cdot 5, 1 \cdot 45)$, which have complex eigen-values, and the points $(1 \cdot 5, 0 \cdot 55)$ and $(1 \cdot 5, 0 \cdot 45)$ which have real eigenvalues. The minimum time trajectories, as shown in Figure 5(b), all ultimately approach the disturbed equilibrium point $(1 \cdot 5, 0 \cdot 45)$. The disturbed equilibrium point $(0 \cdot 5, 1 \cdot 55)$ is actually in the interior of $R(U,N^*)$.

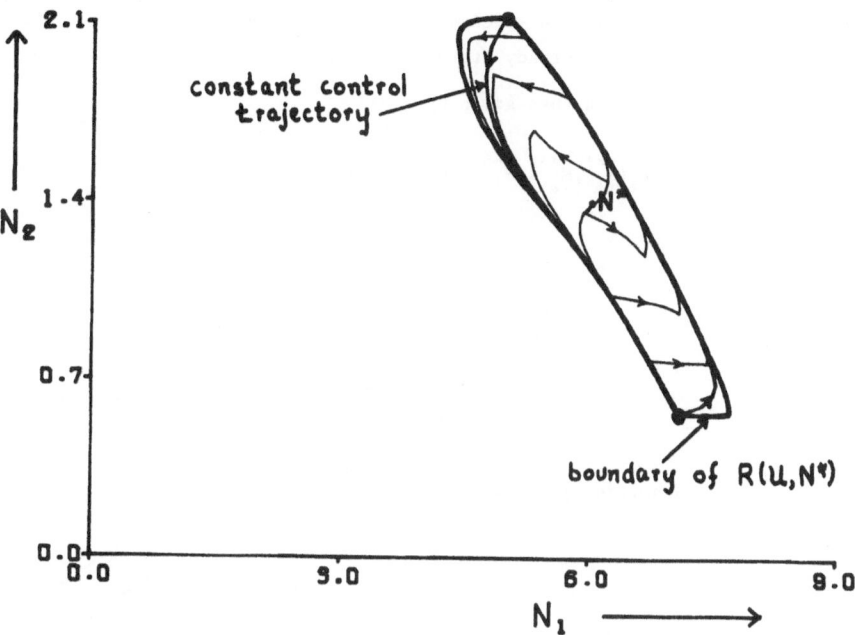

FIGURE 5(a). $R(U,N^*)$ and some constant control trajectories for example 1 with $-0\cdot15 \leq u_i(t) \leq 0\cdot15$

OPTIMAL HARVESTING AND NONVULNERABILITY

In this section, we look at some examples of the effects of optimal harvesting on the nonvulnerability of two species ecosystem models.

Harvesting of both species

Consider the two species ecosystem model

$$\dot{N}_i = f_i(N_1,N_2) - E_iN_i, \quad i = 1,2 , \tag{15}$$

where E_i represents the harvesting effort of the ith species. We will choose the E_i so as to maximise the total revenue over some time interval $[0,L]$. The objective function for this optimal control problem is

$$\int_0^L (v_1E_1N_1 + v_2E_2N_2) \, dt ,$$

where v_i is the price of the ith species. Under optimal harvesting, the system will be driven to an optimal steady state denoted by $N^{**} = (N_1^{**}, N_2^{**})$ and the optimal harvesting effort is then given by

$$E_i = f_i(N_1^{**}, N_2^{**})/N_i^{**} \quad \text{for} \quad i = 1,2 \ . \tag{16}$$

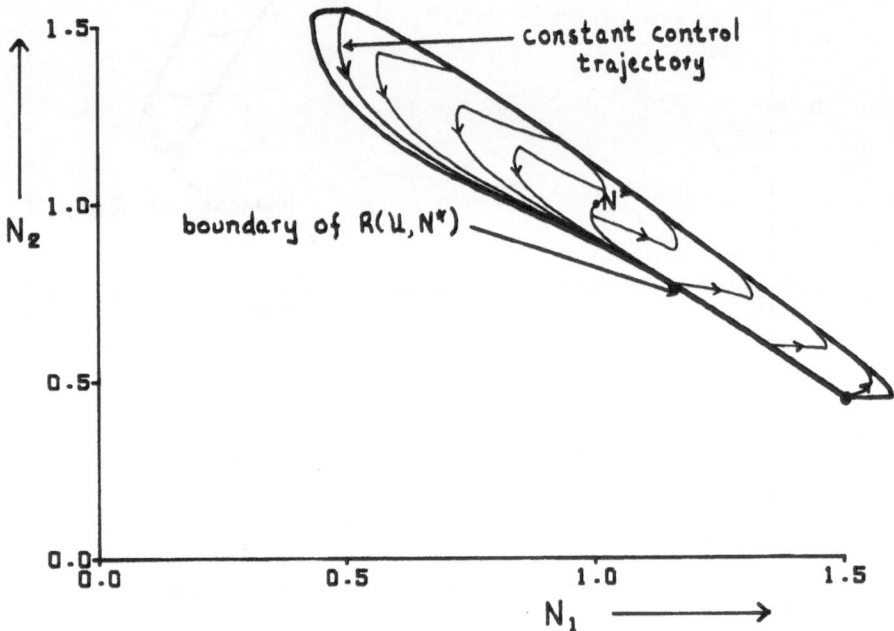

FIGURE 5 (b). $R(U,N^*)$ and some constant control trajectories for example 4

Now introduce disturbances $u(t) \in U$ into system (15), i.e.,

$$\dot{N}_i = f_i(N_1, N_2) - E_i N_i + u_i N_i \ , \quad i = 1,2 \ . \tag{17}$$

For a given set of disturbances U, we can compute the reachable set for N^{**}, i.e., $R(U,N^{**})$. This will enable us to determine, for given sets S and Z, the effect of optimal harvesting on the nonvulnerability of the ecosystem.

Example 5.

$$\dot{N}_1 = N_1(1 - N_1 - 0 \cdot 8N_2) - E_1N_1 + u_1N_1 \; ,$$

$$\dot{N}_2 = N_2(1 - 0 \cdot 25N_1 - N_2) - E_2N_2 + u_2N_2 \; ,$$

with $-0 \cdot 1 \le u_i(t) \le 0 \cdot 1 \; , \quad i = 1,2 \;$.

The undisturbed equilibrium of this competition model is $(0 \cdot 25, \; 0 \cdot 9375)$.

Suppose both species are of equal value. Then under optimal harvesting the equilibrium shifts to the point $N^{**} = (0 \cdot 328, \; 0 \cdot 328)$. From equation (16), the corresponding steady state harvesting policy is $E_1 = 0 \cdot 410$ and $E_2 = 0 \cdot 590$. Figure 6 shows the reachable sets $R(U,N^*)$ and $R(U, \; N^{**})$ for this model. Suppose that Z is the set of all states in which either or both species are depressed to below 15% of the equilibrium values N_1^* and N_2^* . From Figure 6, we see that without harvesting, $R(U,N^*)$ contains states in which the first species population is depressed to below 15% of N_1^* . Hence the system becomes vulnerable over a large enough time interval. With optimal harvesting, the model is nonvulnerable with respect to $\{U,S,Z\}$ for any set S contained in $R(U,N^{**})$. This result is counterintuitive.

Harvesting of one species

Consider the ecosystem model (15) in which only one species is harvested. If we again maximise the total biomass yield, during a large time horizon the system should be driven to a new steady state, N^{**} . If, for example, N_1 is the harvested species the steady state optimal harvesting effort is given by

$$E_1 = f_1(N_1^{**}, \; N_2^{**})/N_1^{**} \; .$$

Example 6. As an illustration, consider the Lotka-Volterra competition model

$$\dot{N}_1 = N_1(1 \cdot 1 - N_1 - 0 \cdot 1N_2) - E_1N_1 + u_1N_1 \; ,$$

$$\dot{N}_2 = N_2(4 - 3N_1 - N_2) + u_2N_2 \; ,$$

in which the first species is harvested. This model has an undisturbed equilibrium point at $N^* = (1,1)$ which is globally stable (Goh, 1976). Under optimal harvesting the new steady state solution is $N^{**} = (0 \cdot 5, \; 2 \cdot 5)$ and the optimal harvesting effort is $E_1 = 0 \cdot 35$. Figure 7 shows the effect of introducing disturbances u(t) into the model, where $-0 \cdot 15 \le u_i(t) \le 0 \cdot 15$, for $i = 1,2$. We see that the introduction of optimal harvesting has, if anything, made the system less vulnerable since $R(U,N^{**})$ is further from the $N_2 = 0$ axis than $R(U,N^*)$.

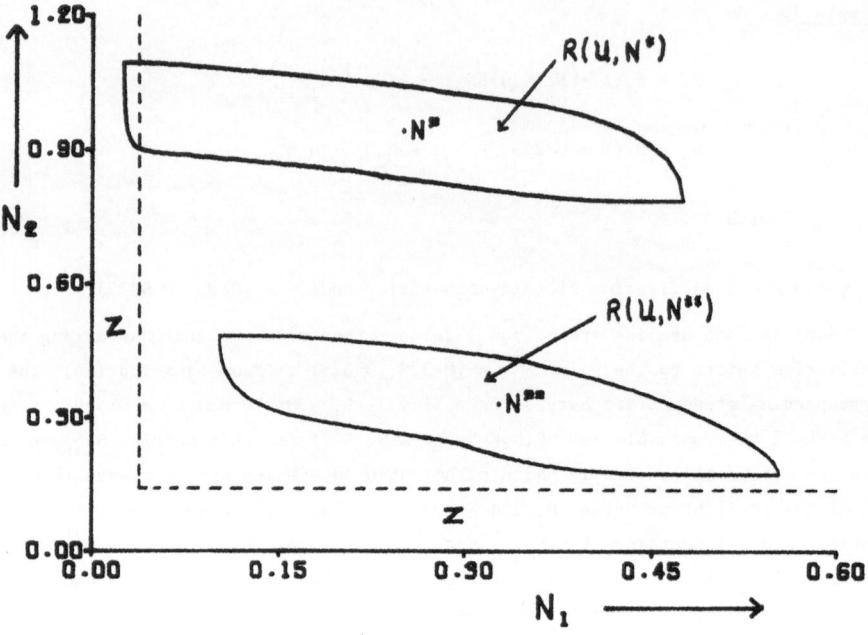

FIGURE 6.   Reachable sets for example 5 with and without 2 species
             optimal harvesting

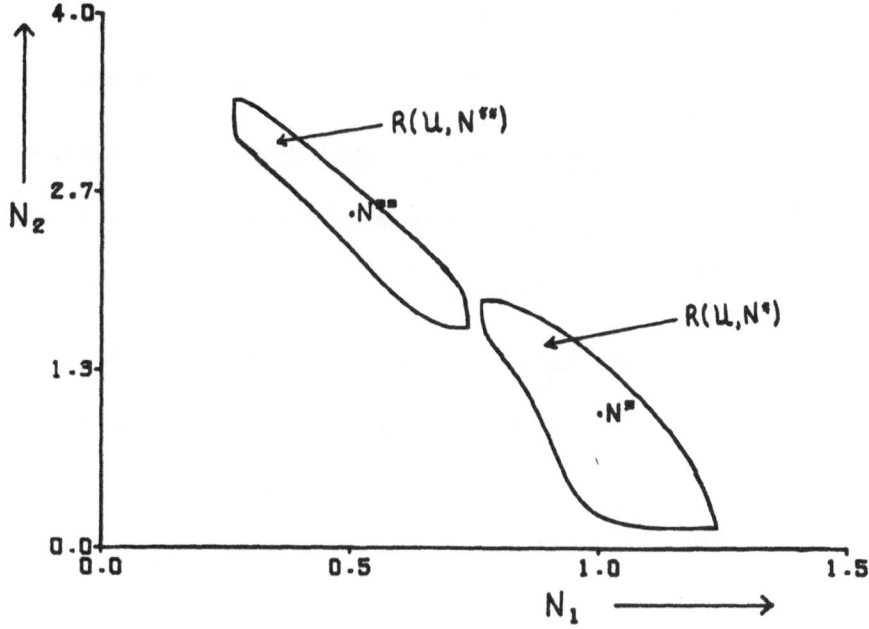

FIGURE 7.   Reachable sets for example 6 with and without optimal
harvesting of N_1

REFERENCES

Gilpin, M.E. and Ayala, F.J. 1973. Global Models of Growth and Competition, Proc. Nat. Acad. Sci. USA, 70, 3590-3593

Goh, B.S. 1975. Stability, Vulnerability and Persistence of Complex Ecosystems, Ecological Modelling, 1, 105-116

Goh, B.S. 1976. Nonvulnerability of Ecosystems in Unpredictable Environments, Theor. Pop. Biol., 10, 83-95

Goh, B.S. 1978. Global Stability in a Class of Prey-Predator Models, Bull. Math. Biol., 40, 525-533

Goh, B.S. 1979. Robust Stability Concepts For Ecosystem Models, in Theoretical Systems Ecology (E. Halfon, Editor). Academic Press

Grantham, W.J. and Vincent, T.L. 1975. A Controllability Maximum Principle, JOTA, 17, 93-114

Hahn, W. 1967. Stability of Motion, Springer-Verlag, Berlin and New York.

Harrison, G.W. 1979. Persistent Sets via Liapunov Functions, Nonlinear Analysis, 3, 73-80

LaSalle, J.P. and Lefschetz, S. 1961. Stability by Liapunov's Direct Method with Applications. Academic Press, New York

Vincent, T.L. and Anderson, L.R. 1979. Return Time and Vulnerability for a Food Chain Model, Theor. Pop. Biol., 15, 217-231

Weiss, L. and Infante, E.F. 1967. Finite Time Stability under Perturbing Forces and on Product Spaces, IEEE Trans. Automatic Control, 12, 54-59

ESTIMATING CONTROLLABILITY BOUNDARIES

FOR UNCERTAIN SYSTEMS

Walter J. Grantham
Department of Mechanical Engineering
Washington State University
Pullman, Washington, USA 99164

Two methods, combining controllability and Lyapunov stability, are
presented for estimating reachability boundaries for nonlinear control systems. The controllability method is exact, but lacks appropriate boundary conditions in certain cases and is effectively
restricted to two-dimensional problems. The approximate Lyapunov
method combines Lyapunov stability theory with a controllability
maximum principle. The Lyapunov estimate of the reachable set generally differs from the actual reachable set, but the estimate is
conservative (i.e., guaranteed to contain the actual reachable set),
does not require integration of the equations of motion, and is
applicable to n-dimensional systems. Both methods are applied to
an example of a prey-predator fishery with bounded harvesting efforts.

INTRODUCTION

Consider the nonlinear control system

$$\dot{x} = f(x,u) \tag{1}$$

where $x \in E^n$ is the state, $u \in U \subseteq E^m$ is the control, and (˙) denotes differentiation with respect to time t. The actual control (or disturbance) $u(t)$ is considered unknown, but the constraint set U is given. The function $f(\cdot): E^n \times E^m \to E^n$
is assumed to be C^1 in x and continuous in u.

A point $x \in E^n$ is a <u>controlled equilibrium point</u> for (1) if there exists a
$\hat{u} \in U$ such that $f(\hat{x},\hat{u}) = 0$. A point x is <u>reachable from</u> $\hat{x}$ if there exists a piecewise continuous control $u(t)$ and a corresponding solution $x(t)$ to (1) which trans-
fers the state from $x(0) = \hat{x}$ to $x(T) = x$ in a finite time interval [0,T], with
$u(t) \in U$ for all $t \in [0,T]$. The <u>reachable set R from</u> $\hat{x}$ is the set of all points
$x \in E^n$ which are reachable from $\hat{x}$. We will be concerned with the case in which R
is not all of E^n. Thus, R has a boundary ∂R, and it is this boundary that we wish
to determine. We restrict the discussion to the case in which R has a nonempty
interior $\overset{\circ}{R}$. The reachable set may contain some, all, or none of its boundary
points. We will be particularly concerned with cases where R is open.

A controllability maximum principle forms the bases for an exact method of determining ∂R, since ∂R generally consists of system trajectories. If the reachable
set is open, this method lacks appropriate boundary conditions. The method may

still be employed however, by allowing the system trajectories to approach ∂R asymptotically. Although the method has been applied to a variety of problems, an explicit guarantee that this asymptotic behavior will actually occur does not currently exist. To overcome this potential difficulty, a similar method (which we will only briefly mention) employs minimum time trajectories, but at the cost of more complex switching structures.

Both the controllability approach (which we will present in more detail) and the minimum time approach are effectively limited to two-dimensional problems since they seek to plot ∂R using system trajectories. In order to handle n-dimensional problems, we will also briefly explore an approximate method based on Lyapunov stability.

As a simple example to illustrate the controllability and Lyapunov methods for estimating reachable sets, we consider a Lotka-Volterra model for a krill-whale (i.e., prey-predator) fishery under bounded harvesting. The nondimensional system equations are given by

$$\dot{x}_1 = x_1(1 - x_1 - \alpha x_2 - u_1) \tag{2}$$

$$\dot{x}_2 = \beta x_2(x_1 - x_2 - u_2) , \tag{3}$$

where x_1 is the krill population, x_2 is the whale population, and u_1 and u_2 are the krill and whale harvesting efforts, respectively. In the general case, the harvesting efforts are bounded, i.e., $\underline{u}_1 \leq u_1 \leq \bar{u}_1$, and $\underline{u}_2 \leq u_2 \leq \bar{u}_2$. For simplicity, we will hold the whale fishing effort constant ($u_2 = 0.2$) and restrict the krill fishing effort to $0 \leq u_1 \leq 0.5$. The other parameter values chosen for our example are $\alpha = 1$ and $\beta = 0.5$.

The topic of vulnerability (Goh 1975, 1976; Vincent and Anderson 1979) is concerned with the following controllability (or reachability) question. Given initial populations (assumed to be a controlled equilibrium point $\hat{x}$) and bounds on the harvesting effort u_1, can either of the species be driven too near to extinction? This question may be answered by determining the reachable set from $\hat{x}$ under all possible harvesting programs and comparing this set with the minimum acceptable population levels.

In this paper, we first present a controllability approach to determining the reachability boundary. Limitations of the approach are discussed and an approximate Lyapunov approach is presented. The approaches are presented in terms of the general system (1) and are applied to the krill-whale system (2)-(3).

A CONTROLLABILITY MAXIMUM PRINCIPLE

Current approaches to determining the reachable set R for (1) seek to generate the reachability boundary ∂R as a locus of system trajectories. In particular, a control $u^*(t)$ is sought which will keep the system in ∂R or which will transfer the system asymptotically to ∂R. Such a control $u^*(t)$ will be termed a boundary control. The controllability approach (Grantham 1973, Grantham and Vincent 1975) uses the "abnormal" case (Vincent and Goh 1972) of Pontryagin's maximum principle (Pontryagin et al., 1962) as a necessary condition for a control to generate a trajectory in ∂R. The minimum time approach (Vincent and Anderson 1979) is similar and seeks trajectories which asymptotically approach ∂R as fast as possible. Because of the strong similarity between the two approaches and because the control switching structure for the minimum time approach is generally more complex, we will concentrate on the controllability approach.

The boundary of the reachable set is an "impermeable surface" (Isaacs 1965) in that no system trajectory can cross or even reach ∂R starting from inside the reachable set R (Grantham 1973). Trajectories may however, approach ∂R asymptotically and may penetrate ∂R from the outside. If two points x(0) and x(T) lie in ∂R for some trajectory x(t), t ε [0,T] then the entire trajectory lies in ∂R. If the reachable set is closed, the existence of such trajectories is ensured. Even if the reachable set is not closed, such trajectories generally exist. This is the case in the krill-whale example in which the reachable set from a positive equilibrium point $\hat{x}$ is an open set whose boundary ∂R is a (stable) limit cycle in the positive quadrant.

To determine the boundary of the reachable set we employ the following controllability maximum principle (Grantham 1973, Grantham and Vincent 1975):

Theorem 1: If $u^*(t)$ ε U generates a trajectory $x^*(t)$ ε ∂R for all t ε [0,T], then there exists a nonzero continuous vector $\lambda(t)$ ε E^n such that for (almost) all t ε [0,T],

$$H[x^*(t),u^*(t),\lambda(t)] = \max_{u \varepsilon U} H[x^*(t),u,\lambda(t)] = 0 \tag{4}$$

$$\dot{\lambda}^T = - \frac{\partial H[x^*(t),u^*(t), \lambda]}{\partial x} \tag{5}$$

where

$$H(x,u,\lambda) \overset{\Delta}{=} \lambda^T f(x,u) . \tag{6}$$

If the reachable set R is locally (near a point $x^*(t)$ ε ∂R) of the form

$$\overset{\circ}{R} = \{x \ \epsilon \ E^n \mid g(x) < 0\} \tag{7}$$

$$\partial R = \{x \ \epsilon \ E^n \mid g(x) = 0\} \tag{8}$$

where $g(\cdot)$: $E^n \to E^1$ is C^1 with nonzero gradient, then

$$\lambda^T(t) = k(t) \frac{\partial g[x^*(t)]}{\partial x} \tag{9}$$

where $k(t)$ is a positive scalar.

Proof: (See Grantham 1973).

Note that the controllability maximum principle of Theorem 1 differs from the optimal control maximum principle in that no explicit performance measure is employed, such as time or cost. Theorem 1 is essentially a geometric condition on the set of "velocity" vectors $\dot{x} = f(x,u)$, $u \ \epsilon \ U$. The maximum condition (4) reflects the inability to penetrate ∂R from the inside. The condition $H = 0$ corresponds to the boundary trajectory being tangent to ∂R (i.e., contained in ∂R). The interpretation (9) and the absolute continuity of $\lambda(t)$, which follows from (5), implies that no trajectory in ∂R can cross a "corner" in ∂R and remain in ∂R. As we shall see, such corners can occur when ∂R contains an equilibrium point.

To apply Theorem 1 to the krill-whale example, we form

$$H = \lambda_1 \dot{x}_1 + \lambda_2 \dot{x}_2 \ , \tag{10}$$

that is,

$$H = \lambda_1 x_1 (1 - x_1 - \alpha x_2 - u_1) + \lambda_2 \ \beta x_2 (x_1 - x_2 - u_2) \ . \tag{11}$$

Focusing on the positive quadrant ($x_1 > 0$, $x_2 > 0$), considering u_2 as a constant ($u_2 = 0.2$), and applying (4) with $0 \leq u_1 \leq 0.5$, we have

$$u_1^*(t) = \begin{cases} 0.5 & \text{if } \lambda_1(t) < 0 \\ \\ 0 & \text{if } \lambda_1(t) > 0 \ . \end{cases} \tag{12}$$

The singular case $\lambda_1(t) \equiv 0$ cannot occur. To see this we note that the adjoint equations (5) are

$$\dot{\lambda}_1 = - \lambda_1 [1 - u_1 - 2x_1 - \alpha x_2] - \lambda_2 \beta x_2 \tag{13}$$

$$\dot{\lambda}_2 = \lambda_1 \alpha x_1 - \lambda_2 \beta [x_1 - 2x_2 - u_2] \ . \tag{14}$$

Thus, $\lambda_1(t) \equiv 0$ and $x_2(t) > 0$ would require $\lambda_2(t) \equiv 0$ and would therefore, contradict the contition $\lambda \neq 0$ in Theorem 1.

Thus, we see that boundary control $u_1^*(t)$ is bang-bang, switching between no krill harvesting and maximum harvesting as the state $x^*(t)$ crosses certain switching arcs. For minimum time (or other optimal) control the corresponding switching arcs may be quite complex and difficult to obtain even for two-dimensional systems. With the controllability maximum principle (Theorem 1) the switching arcs are more easily obtained.

Since switching occurs when $\lambda_1(t) = 0$ and since $H = 0$ with $\lambda_2(t) \neq 0$, the switching arc for the krill-whale system is simply the whale isocline $\dot{x}_2 = 0$, as seen from (10). From an examination of the adjoint equations (13) and (14) (immediately before and after a switch) we have

$$u_1^*(x) = \begin{cases} 0.5 \text{ if } \dot{x}_2(x) < 0 \\ \\ 0 \quad \text{if } \dot{x}_2(x) > 0 \ . \end{cases} \tag{15}$$

Since $\dot{x}_2$ depends only on the state x (for fixed u_2) we have the boundary control u_1^* synthesized in a feedback form and the adjoint equations play no further role. In more general circumstances, however, the adjoint equations (5) must be integrated along with the state equations (1) in order to determine switching times.

At this point, a difficulty in applying the controllability maximum principle becomes apparent. We have completely synthesized the control u_1^* necessary to keep the trajectory on ∂R, assuming the system is initially on ∂R. However, for the krill-whale example, the reachable set is open and we have no boundary conditions for determining an initial point $x^0 \in \partial R$, except $H = 0$. For a large class of problems, however, initial conditions do exist and we present these conditions for future reference.

In many instances, the reachable set is closed, for system (1) starting from a specified initial set $\theta \subset E^n$. That is, every point in ∂R is reachable. But no trajectory can reach ∂R from inside $\overset{\circ}{R}$. Thus, there must exist trajectories in ∂R and ∂R must intersect the boundary $\partial \theta$ of the initial set (at a point where $H = 0$). If the initial set θ is a single point ($\theta = \partial \theta = x^0$) and if the reachable set is closed (and not all of E^n) then $x^0 \in \partial R$ and the initial condition problem is solved. On the other hand, if the reachable set is closed and has a nonempty interior, and if the initial set θ is the closure of an open set with a smooth boundary, e.g.,

$$\theta = \{x \in E^n \mid h(x) \leq 0\} \tag{16}$$

where $h(\cdot)$: $E^n \to E^1$ is C^1 with nonzero gradient, then it can be shown (Grantham 1973) that the following transversality conditions must hold at a point x^0 where ∂R intersects $\partial\theta$:

$$h(x^0) = 0 \tag{17}$$

$$\lambda^T_{(0)} = \mu \frac{\partial h(x^0)}{\partial x} \qquad \mu > 0 \; . \tag{18}$$

That is, ∂R is tangent to $\partial\theta$ at x^0 and $\lambda(0)$ is an outward normal to $\partial\theta$. These conditions, along with $H = 0$, provide the needed boundary conditions, providing the reachable set actually is closed.

For the krill-whale example, however, the reachable set starting from any positive equilibrium point $\hat{x}$ is an open set. The boundary ∂R is not itself reachable but could be generated by the control $u_1^*(x)$ if we could find an initial point on ∂R. One approach to this problem is based on the fact that any controlled equilibrium point $\hat{x}$ satisfies $H = 0$ (since $\dot{x} = 0$) and is therefore a candidate for a point on ∂R. The equilibrium points for the krill-whale system under constant controls $\hat{u}_1$ and $\hat{u}_2$ are

$$\hat{x}_1 = \frac{1 - \hat{u}_1 + \alpha\hat{u}_2}{1 + \alpha} \tag{19}$$

$$\hat{x}_2 = \frac{1 - \hat{u}_1 - \hat{u}_2}{1 + \alpha} \; . \tag{20}$$

For $\hat{u}_2 = 0.2$, $\alpha = 1$ and $0 \le \hat{u}_1 \le 0.5$, the possible constant-control equilibrium points lie on the $\dot{x}_2 = 0$ whale isocline, on the line segment from $(x_1, x_2) = (0.35, 0.15)$ to $(0.6, 0.4)$. For any constant value of $\hat{u}_1$, with $0 \le \hat{u}_1 \le 0.5$, and for $\beta = 0.5$ the resulting equilibrium point is a stable focus, with the constant-control trajectories spiraling in about the equilibrium point. Figure 1 illustrates the trajectories under the constant controls $u_1 = 0$ and $u_1 = 0.5$. The region bounded by the cross-hatched curve in Figure 1 is the reachable set R starting from any point inside the set. The reachability boundary is a limit cycle, with the boundary control u_1^* switching between 0 and 0.5 as the trajectory crosses $\dot{x}_2 = 0$. Before indicating more about how this reachability boundary is determined, we continue the discussion of equilibrium points.

At first glance, it might appear from Figure 1 that the krill-whale equilibrium points $\hat{x} = (0.35, 0.15)$ for $\hat{u}_1 = 0.5$ and $\hat{x} = (0.6, 0.4)$ for $\hat{u}_1 = 0$ lie on ∂R. This is not the case. Since each of the two equilibrium points is a focus (for $\beta = 0.5$), a trajectory in ∂R will necessarily spiral around the associated equilibrium point and cross the switching arc $\dot{x}_2 = 0$. Thus, the equilibrium points do not lie on ∂R and we are back to the original initial condition problem.

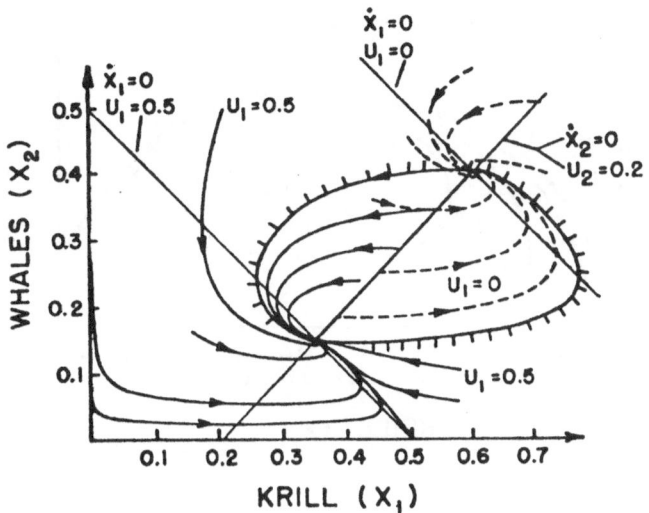

FIGURE 1. Trajectories and Reachable Set for $0 \leq u_1 \leq 0.5$ and $u_2 = 0.2$

Lacking knowledge of a point x^0 precisely on ∂R, Vincent and Anderson (1979) suggest choosing a point x^0 (hopefully inside $\overset{\circ}{R}$ and sufficiently near ∂R) and then using the boundary control u_1^* from Theorem 1. Hopefully, the resulting trajectory will converge to ∂R since the control corresponds to a maximum effort to penetrate ∂R. However, no guarantee exists that this asymptotic behavior will actually occur, starting from an equilibrium point $\hat{x}$. The same idea (of reaching ∂R asymtotically) forms the basis for the previously mentioned minimum time approach. Both approaches are due to Vincent and have been applied to a wide variety of two-dimensional systems (see Vincent and Anderson 1979).

The reachability boundaries shown in Figure 2 (for $\alpha = 1$, $\beta = 0.5$, and various upper bounds on krill harvesting effort) were generated asymptotically using the boundary control u_1^* (x) from Theorem 1. Note that the higher allowable fishing efforts produce a corner in ∂R. Evidently the equilibrium point corresponding to maximum effort harvesting changes from a focus to a node.

A LYAPUNOV APPROACH

The direct application of Theorem 1 employs actual system trajectories to

158

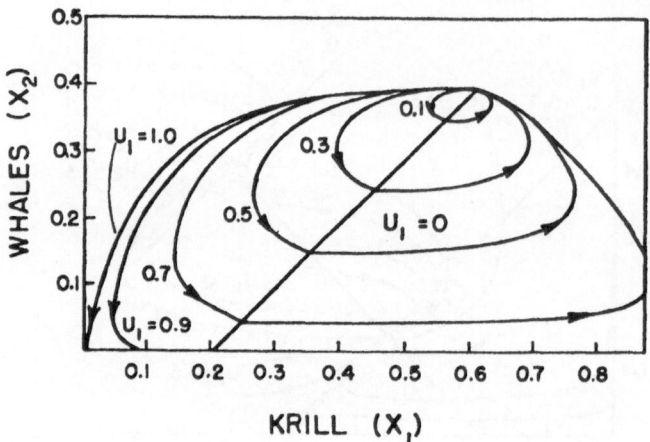

FIGURE 2. Reachable Set Boundaries for Various Upper Bounds on Krill Harvesting
Effort

generate ∂R, the boundary of the reachable set R. The procedure is therefore effectively limited to two-dimensional problems, although there is no conceptual
difficulty in going to higher dimensions. We have also seen then an additional
difficulty occurs when the reachable set is open, since appropriate boundary conditions are no longer available for use with Theorem 1.

To overcome these two difficulties, we present an approximate method for estimating R which combines Theorem 1 with Lyapunov stability theory. The resulting
estimate for the reachable set R generally differs from the actual reachable set.
However the estimate is conservative (it contains R), does not require trajectory
integration, and is applicable to n-dimensional problems.

The basis for the Lyapunov controllability method is given by

Theorem 2: If there exists a C^1 function $V(\cdot): E^n \to E^1$ and a number V^* such
that

$$V(\hat{x}) \leq V^* ,$$ (21)

$$\{x \; \varepsilon \; E^n \; | \; V(x) \leq V^*\} \; \mathbf{C} \; \{x \; \varepsilon \; E^n \; | \; V(x) \leq k\}$$ (22)
$$\text{for all } k > V^* ,$$

and

$$\{x \mid \dot{V}(x,u) \geq 0, \ u \ \varepsilon \ U\} \ \underline{\subseteq} \ \{x \mid V(x) \leq v^*\} \ , \tag{23}$$

then for system (1) the reachable set R from the controlled equilibrium point $\hat{x}$
satisfies

$$R \ \underline{\subseteq} \ \{x \ \varepsilon \ E^n \mid V(x) \leq v^*\} \ , \tag{24}$$

where

$$\dot{V}(x,u) = \frac{\partial V(x)}{\partial x} \ f(x,u) \quad . \tag{25}$$

Proof: From (21) and (22) the initial point $\hat{x}$ is in the region $V(x) \leq v^*$
and this region is inside $V(x) \leq k$ for $k > v^*$. Thus, to leave $V(x) \leq v^*$ the system
must be able to penetrate the surface $V(x) = v^*$ in the direction of increasing $V(x)$.
But this cannot happen since, from (23), $\dot{V}(x,u) < 0$ for all x such that $V(x) > v^*$.
This completes the proof.

The idea behind using Theorem to approximate ∂R is a belief, based on the converse theorems of Lyapunov stability (see Hahn 1967), that for a suitable choice of
the function $V(\cdot)$ satisfying Theorem 1 (called a Lyapunov function), the reachability boundary will coincide with the surface $V(x) = v^*$, for some v^*. Along such a
surface $\lambda(t)$, from Theorem 1, is a positive multiple of $\partial V(x)/\partial x$ and the control
$u^*(x)$ from Theorem 1 is the control which maximizes $\dot{V}(x,u)$. For any given choice
of $V(\cdot)$, however, ∂R and the surface $V(x) = v^*$ will generally not coincide and
$u^*(x)$ will not be the same as the control which maximizes $\dot{V}(x,u)$.

Resolution of these differences will be left to future research. In the mean-
time we employ Theorem 2 as follows. Given a selected function $V(x)$ satisfying
(21) and (22), we compute $\dot{V}(x,u)$ with $u = u^*(x)$ from Theorem 1. If $V(x)$ is such
that the region $\dot{V}[x,u^*(x)] \geq 0$ is contained inside the region $V(x) \leq v^*$ then so is
the reachable set from $\hat{x}$.

For the krill-whale example we will employ three separate Lyapunov functions
to estimate the reachable set. We begin with the Volterra-type function

$$V(x) = \sum_{i=1}^{2} d_i \ [x_i - \hat{x}_i - \hat{x}_i \ln(x_i/\hat{x}_i)] \ , \tag{26}$$

where $\hat{x}$ is the equilibrium point which would result if the harvesting efforts were
held constant. We choose $\hat{x} = (0.6, 0.4)$, corresponding to $\hat{u} = (\hat{u}_1, \hat{u}_2) = (0, 0.2)$.
For $u_2 = 0.2$ we have

$$\dot{V}(x,u_1^*) = -d_1(x_1 - \hat{x}_1)^2 - (d_1\alpha - d_2\beta)(x_1 - \hat{x}_1)(x_2 - \hat{x}_2)$$
$$-d_2\beta(x_2 - \hat{x}_2)^2 - d_1(x_1 - \hat{x}_1)u_1^* \quad . \tag{27}$$

For demonstration purposes we choose $d_1 = 1$ and $d_2 = 2$ so that the cross-product term cancels in (27), with $\alpha = 1$ and $\beta = 0.5$. Therefore, for $u_1 = 0$ we have $\dot{V} \leq 0$ throughout the positive quadrant. For $u_1 = 0.5$ we have $\dot{V} \geq 0$ inside the circular region $(x_1 - 0.35)^2 + (x_2 - 0.4)^2 \leq 0.0625$ (see Figure 3). The largest value of $V(x)$ in this circular region is $V = 0.575$, which occurs at $x^* = (0.1, 0.4)$.

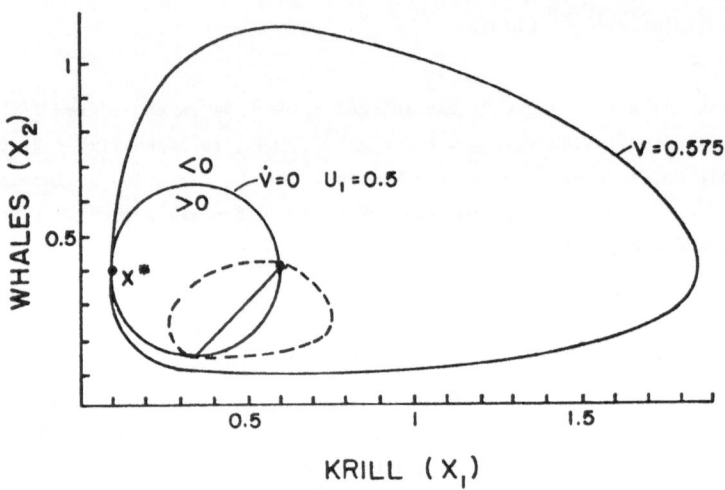

FIGURE 3. Initial Lyapunov Estimate for the Reachable Set

Thus, from Theorem 2 the reachable set R (starting from any positive constant-control equilibrium point) is contained in the region $V(x) \leq V^* = 0.575$. This region conservatively over-estimates the actual reachable set (dashed boundary in Figure 3). A better fit undoubtedly could be found by an alternate selection of parameters in $V(x)$ or by a different choice of $V(x)$.

The estimate $V(x) \leq 0.575$ illustrated in Figure 3 can be improved without altering our choice of $V(x)$ by using additional Lyapunov functions to reduce the region, as illustrated in Figure 4. For $u_1 = 0$ (i.e., below the $\dot{x}_2 = 0$ isocline) the function $V_1(x) = x_1$ is a Lyapunov function with $\dot{V}_1(x, u_1^*) < 0$ to the right of the $\dot{x}_1 = 0$, $u_1 = 0$ krill isocline ($1 - x_1 - x_2 = 0$). Thus the actual reachable set cannot contain points to the right of the vertical line $x_1 = 0.901$ through the point of intersection of the curve $V(x) = 0.575$ and the line $1 - x_1 - x_2 = 0$. Similarly, the function $V_2(x) = x_2$ is a Lyapunov function with $\dot{V}_2(x, u_1^*) < 0$ above the whale isocline. Thus, the reachable set cannot contain points above the horizontal line

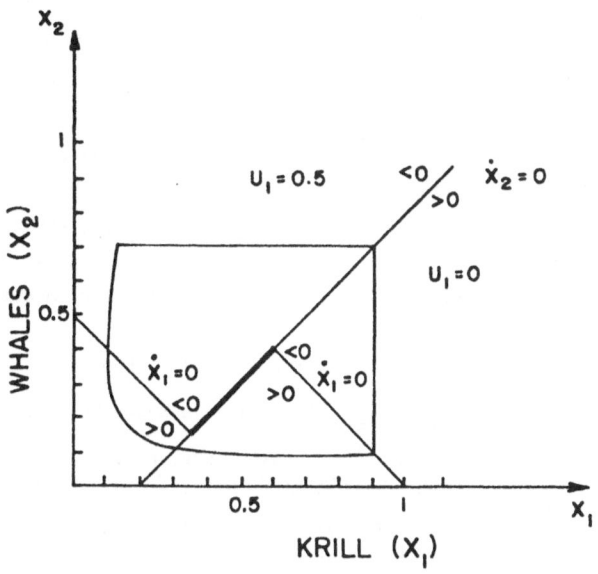

FIGURE 4. Reduced Lyapunov Estimate

$x_2 = 0.701$ which passes through the point of intersection of the line $x_1 = 0.901$ and the whale isocline $x_1 - x_2 = 0.2$.

SUMMARY

Two methods have been discussed for estimating the reachable set R from a controlled equilibrium point. The controllability approach is exact but lacks appropriate boundary conditions when R is open. In addition, the method is effectively limited to two-dimensional problems. An approximate method, combining controllability and Lyapunov stability, has also been presented. This method is conservative, does not require trajectory integration and is applicable to n-dimensional problems.

REFERENCES

Goh, B. S. 1975. Stability, Vulnerability, and Persistence of Complex Ecosystems, Ecological Modeling, 1, 105-116

Goh, B. S. 1976. Nonvulnerability of Ecosystems in Unpredictable Environments, Theoretical Population Biology, 10, 83-95

Grantham, W. J. 1973. A Controllability Minimum Principle, PhD Dissertation, University of Arizona, Tucson

Grantham, W. J. and Vincent, T. L., 1975. A Controllability Minimum Principle, J. of Optimization Theory and Applications, 17, 93-114

Hahn, W. 1967. Stability of Motion, Springer-Verlag, New York

Isaacs, R. 1965. Differential Games, John Wiley & Sons, New York

Pontryagin, L. S., Boltyanskii, V. G., Gamkrelidze, R. V., and Mischenko, E. F. 1962. The Mathematical Theory of Optimal Processes, John Wiley & Sons, New York

Vincent, T. L. and Anderson, L. R. 1979. Return Time and Vulnerability for a Food Chain Model, Theoretical Population Biology, 15, 217-231

Vincent, T. L. and Goh, B. S. 1972. Terminality, Normality, and Transversality Conditions, J. of Optimization Theory and Applications, 9, 32-50

ESTIMATING THE EFFECT OF KRILL HARVESTING ON THE SOUTHERN OCEAN ECOSYSTEM

J. R. Beddington
Biology Department, University of York
Heslington, York

The development of a fishery for krill and the recent signing of the Antarctic Convention on the Conservation of Living Marine Resources, which is aimed at controlling exploitation within the Southern Ocean, sets the context for the problem addressed in this paper. That problem is to determine the level of krill harvesting that will permit the recovery of the baleen whales to some reasonable proportion of their initial abundance. Tentative solutions of this problem are described in this paper in terms of simple empirical models. These models may be viewed as providing some empirical underpinning of the theoretical considerations of May et al. (1979).

INTRODUCTION

Starting with the establishment of a land station in South Georgia in 1908 the baleen whales of the Antarctic have been harvested throughout the century. Major exploitation began with the establishment of pelagic fleets in the early 1930s and although much reduced continues today. The pattern of exploitation has been largely determined by economic considerations. Easily caught humpback whales (Megaptera novaeangliae) were the first species to be significantly depleted; blue whales (Balaenoptera musculus), the largest and most profitable, were depleted next and the industry then switched to fin whales (Balaenoptera physalus). Depletion of these and the change in emphasis of the industry from oil-based to meat-based production led to exploitation of the sei whale (Balaenoptera borealis). Finally the commercial exploitation of the minke whale (Balaenoptera acutorostrata), the smallest of the baleen whales, started in 1970 and is currently the only species still exploited in the Antarctic.

Although there are problems in identifying separate stocks or populations of baleen whales, the Scientific Committee of the International Whaling Commission has defined six areas which are assumed to be relatively discrete. These areas illustrated in Fig. 1 and discussed by Mackintosh (1965) have experienced rather different historical patterns of exploitation and any reasonable modelling of the system needs to take this into consideration. The progressive depletion of the whale species has been accompanied by remarkably large changes in the demographic parameters. Pregnancy rates and ages of maturation has apparently responded to increased availability of the common food resource krill, Euphausia superba.

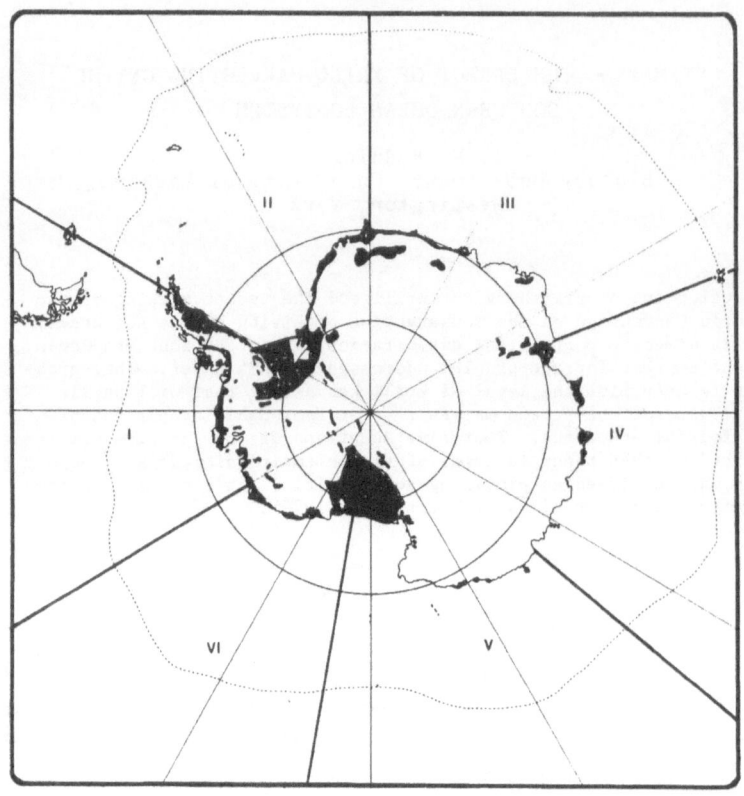

FIGURE 1. Antarctica and the Southern Ocean: IWC Whaling Areas 1 - VI

ESTIMATION OF WHALE POPULATIONS

The demographic models used in whale estimation may usefully be written in matrix vector notation.

$$V_{t+1} = M\ V_t \tag{1}$$

where V_t is a vector of age and sex classes and M is a Leslie-type transition matrix whose elements are time variant. At equlibrium, presumed to occur prior to exploitation, the equilibrium vector V is given by

$$V = M\ V \tag{2}$$

This equation then is used to determine the relationship between demographic parameters prior to exploitation. Absolute values of the vector V are sought in the estimation procedure. With exploitation the model is extended so that

$$V_{t+1} = M V_t - C_t \qquad (3)$$

where C_t is the vector of catches by age and sex in year t.

Population estimation using this dynamic model consists in finding the vector V that either minimizes the function

$$\Psi(V) = \sum_t (C_t - P_t)^2$$

where C_t is the observed catch and P_t that expected for known effort in year t, or the function

$$\Psi(V) = \sum_t (S_t - E_t)^2$$

where S_t is the number of sightings and E_t that expected for known sighting effort.

The calibration of whaling effort is a tendentious process and quite recently significant biases have been identified in the measure of effort used by the IWC (Beddington (1979) and Kirkwood (1979)) which rendered invalid the implicit assumption in the above analysis that catch per unit effort was proportional to population size. The current measure of effort used by the IWC and in this study is catcher boat searching hours.

Once an initial population size has been estimated the underlying demographic model and catch history produces a population trajectory to the present day. Figure 2 illustrates the population trajectories of the exploited whale species in the S.E. Indian Ocean, Area (IV). The result illustrated may be taken as typical of other areas of the Southern Ocean with the caveat that the blue whale data in Area IV are poor and that the initial population abundance may be somewhat overestimated. In the Southern Ocean as a whole, fin whale abundance was substantially higher than that of blue prior to exploitation of either species.

ESTIMATION OF CHANGES IN KRILL CONSUMPTION WITH BALEEN WHALE EXPLOITATION

The population trajectories estimated for the baleen whale species provide a basis for estimating changes in the consumption of krill. The details of the calculations involved will be published elsewhere (Beddington 1980) and only a brief outline is given here. The population numbers are converted by means of age - length - weight relationships to biomass. Estimates of energetic requirements of whales to satisfy resting metabolism and provide for growth and movement were then made.

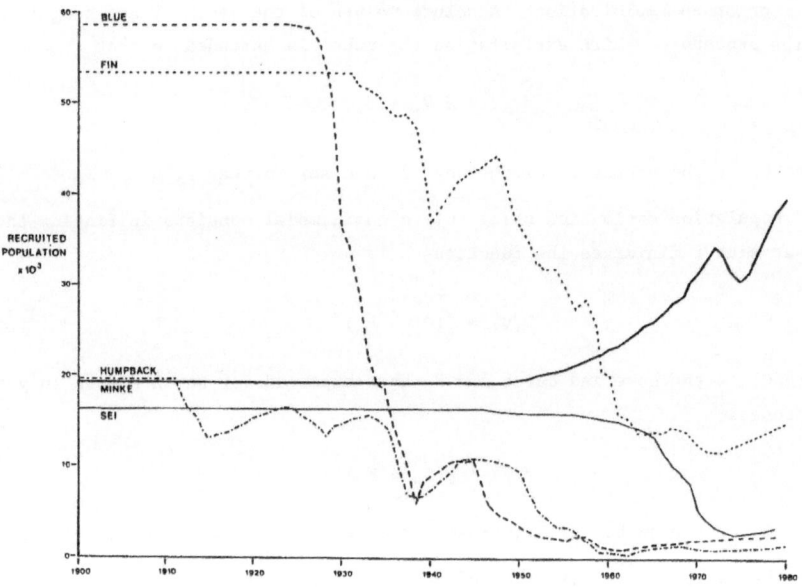

FIGURE 2. Estimates of Population Changes for Baleen Whales, Antarctic Area IV,
1900 - 1980

These permitted, with knowledge of the diet and calorific content of the diet, estimates to be made of the amount of krill eaten by each whale species over the
period from 1900 to the present. Using this technique calculations were made of
the changes in krill consumption by the baleen whales from the pristine state of
the system through to the current state. For area IV these are summarised in
Table 1.

TABLE 1

	Krill consumption by baleen whales Area IV		
	1900	1970	1980
Consumption x 10^6 metric tonnes	26.4	3.4	4.8

These consumption trajectories are the basis of the analysis of the subsequent section.

DEMOGRAPHIC CHANGE AND KRILL AVAILABILITY

The change in demographic parameters that has occurred within Area IV for the three species for which data are available is illustrated in Figure 3.

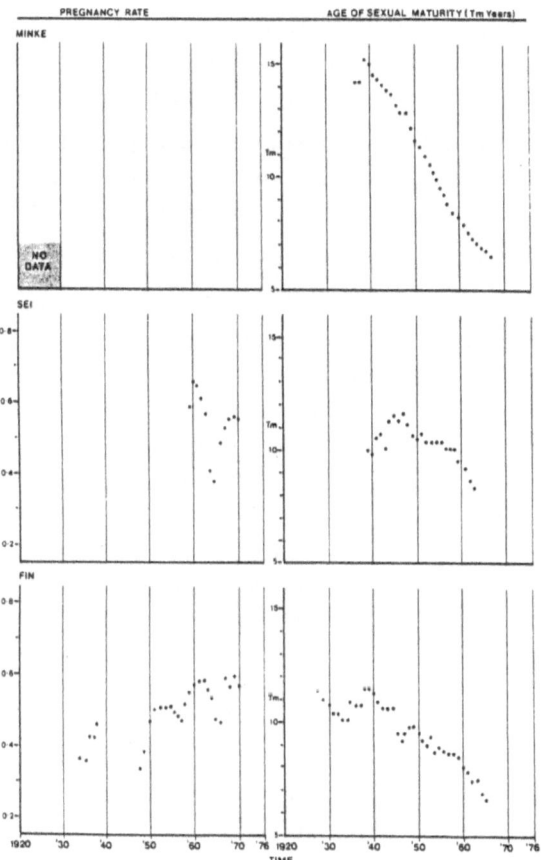

FIGURE 3. Demographic Data: AREA 4.   Minke, Sei and Fin Whales

The picture of improved demographic response is clear. Particularly striking is the change in age at sexual naturity of the minke whale which has dropped almost 10 years. There is a good evidence to indicate that maturity occurs at roughly a constant size, hence the change is likely to be determined not by social factors, but by an increased growth rate. A full analysis of the calculation that went into these presented figures may be found in Free and Beddington (1980).

The increased availability of krill implied by this response is likely to have occurred in two ways as the level of predation has decreased. These are illustrated in Figure 4. Although it may be expected that the demographic response would saturate at high food availability as a reasonable first approximation it was decided to use a linear model relating changing consumption to the changing demographic variables, consumption being assumed to be negatively correlated with krill abundance[*] by the two mechanisms illustrated in Figure 4.

The pattern of residuals should then show any important deviation from linearity. These residuals if inspected against time should also indicate changes in krill recruitment and/or changes in predation levels of other krill-eating species.

The results for Area IV are summarised in Table 2 and illustrated in Figure 5.

Although the analysis presented here is for one area of the Southern Ocean, similar results have been obtained for other areas and they are notable. Firstly, the linear models work well and there is no evidence from the basic residuals that non-linearity in the response is having an important effect. (The exception is the sei whale pregnancy rate, where it is clear that a response has already occurred prior to there being any data available.) Secondly, the residual patterns show no time trends that might indicate a buildup of competitive species like crabeater seals, cephalopods or penguins. Thirdly, there appears to be a pronounced oscillation with time of the residuals. This is more pronounced for the pregnancy rate data. This may reflect the shorter period of krill abundance important in the determination of pregnancy rates. Among a number of alternative explanations, one that seems worthy of investigation is that this oscillation reflects changes in krill abundance. However, if for the moment this effect is ignored, then these empirical models when coupled with demographic models offer the possiblility for answering the question posed earlier: namely, what are the likely effects of a krill fishery on baleen whale recovery?

---

[*] An important point of detail in the analysis is that the krill abundance in the preceding year will affect pregnancy rate. By contrast the age of sexual maturity of a whale will be determined by the abundance from birth to sexual maturity. Appropriate time lagged and averaged levels of consumption were therefore used in the fitting procedures.

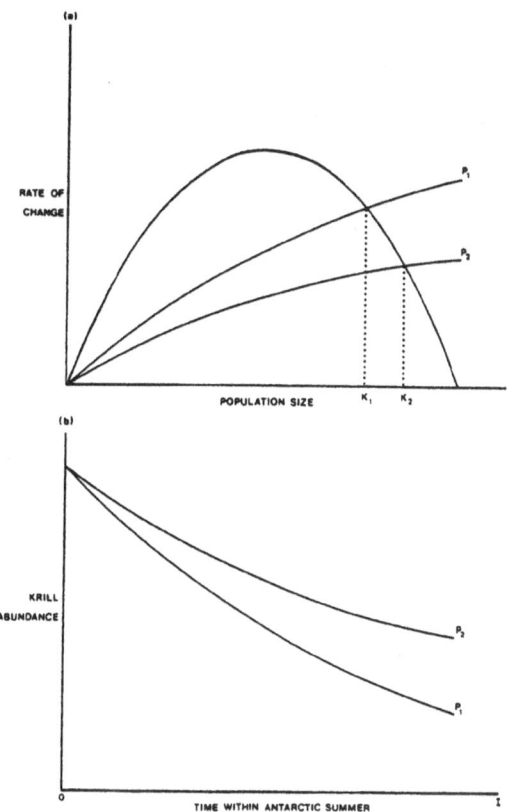

FIGURE 4. P_1 and P_2 are predation rates at high and low population sizes of baleen whales. Figure (a) illustrates the change in overall equilibrium population of the krill, including recruitment, from K_1 at high predation levels to K_2 at low predation levels. Figure (b) illustrates the change in krill abundance within the summer feeding period in the Antarctic.

The operation of a krill fishery is likely to be very similar to the baleen whale feeding patterns, in concentrating on the large krill swarms. Hence predictive models may be constructed using equations of type (1) and (3) in which the demographic parameters are determined by the level of consumption including fishing of krill. The underlying assumption is that other krill-eating species are sufficiently dissimilar to the baleen whales to have only a marginal effect on their dynamics. If these rather strong assumptions are allowed the equilibrium of the baleen whale/krill system will be reached when krill consumption reaches its

TABLE 2

Summary of Statistical Analysis for Area IV

Species	Age of Sexual Maturity					'True' Pregnancy Rate				
	Years	Regression Coefficient	Constant Term	Coefficient of Determination	F Value (d,f)	Years*	Regression Coefficient	Constant Term	Coefficient of Determination	F Value (d,f)
Fin	1928-1965	.182	7.524	.623	59.4(1,36)	1934-1970	-.0057	.372	.607	40.2(1,26)
Sei	1939-1963	.232	8.386	.654	43.5(1,23)	1958-1969	No significant fit			
Minke	1933-1966	.630	5.760	.844	178.7(1,33)	Insufficient data for analysis				

* Excluding war years 1939-1946

REGRESSION MODELS

$Tm(t) = a \bar{C} + b,$ $F(t) = a C(t-1) + b$

$Tm(t) \equiv$ Age of sexual maturity of a cohort born in year (t).

$\bar{C} \equiv$ Average consumption in period $t - t + Tm(t)$

$F(t) \equiv$ Pregnancy rate

$C(t) \equiv$ Consumption in year t.

pristine level. This implies that both the pattern of whale harvesting and the time path of the krill fishery development will significantly affect the equilibrium species configuration of the baleen whales. It is here that an intriguing control theoretic problem may be posed.

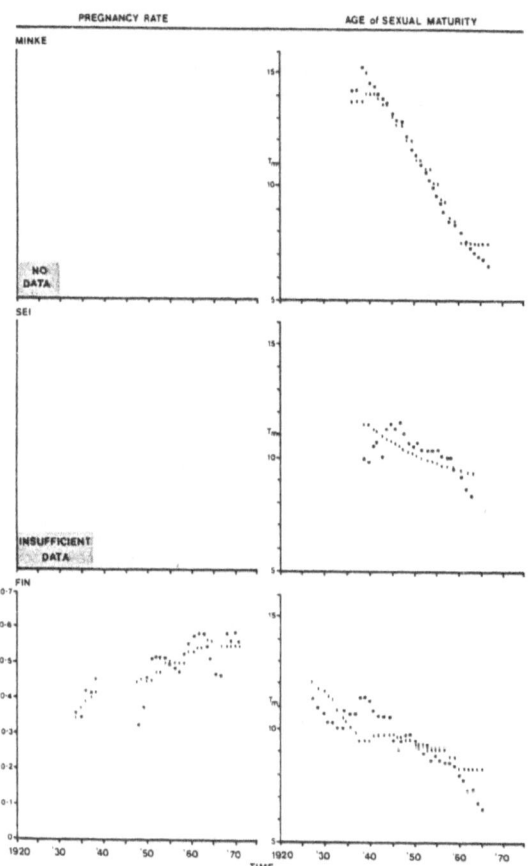

FIGURE 5. Fitted Relationships of Demographic Parameters against Krill
Consumption over Time.   AREA 4
• - observed ǀ - expected

PREDICTING WHALE DYNAMICS

Formally for each whale species there is a model of Form (3)

$$V_{t+1} = M V_t - C_t$$

in which the elements of M are determined in part endogenously by the consumption of the whale species itself at each time period and in part exogenously by the consumption of other species and of the level of the krill fishery. Projected catches by sex and age in year t are contained in the vector C_t .

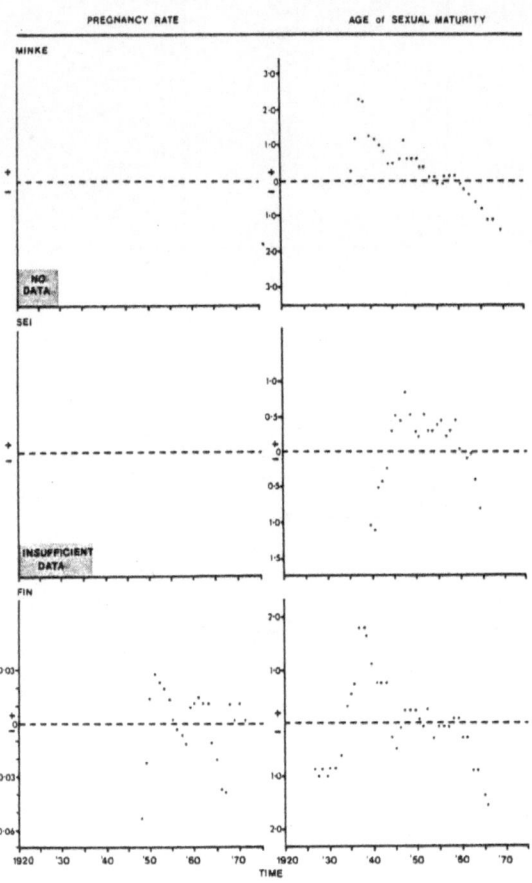

FIGURE 5.   Residuals over Time.   AREA 4

The problem is to choose for the set of such models the levels of krill harvesting and whale harvesting that will bring the system as quickly as possible to some specified equilibrium configuration.

This problem may most readily be solved using simulation, however there are practical problems involved in the implementation of a solution. There is currently a ban on all pelagic whaling in the Antarctic save that for Minke whales. Whale quotas are set by the International Whaling Commission, krill quotas will be set by a commission formed by the signatories to the convention for the Conservation of Antarctic Living Marine Resources. Furthermore, arguments on whether the ideal species configuration of the ecosystem is one with many whales and a low krill harvest or a high krill harvest and few whales have yet to be resolved.

REFERENCES

Beddington, J.R. 1979. On Some Problems of Estimating Population Abundance from Catch Data. Rep. Int. Whal. Comm. 29, pp 149-154

Beddington, J.R. 1980. The Modelling and Management of the Southern Ocean. IUCN (to be published)

Everson, I. 1977. The Living Resources of the Southern Ocean. FAO, Southern Ocean Fisheries Survey Programme GLO/SO/77/1: 155 p.

Free, C.A. and Beddington, J.R. 1980. A technique for estimating the mean and variance of the distribution of the age of sexual maturity for data obtained from the transition phase of earplug laminae. Rep. Int. Whal. Comm. 30 (In press)

Kirkwood, G.P. 1979. The Net Catcher Day as a Measure of Effort. Rep. Int. Whal. Comm. 29, pp 163-166

Mackintosh, N.A. 1965. The Stocks of Whales Fishing News (Books) Ltd.

May, R.M., Beddington, J.R., Clark, C.W., Holt, S.J. and Laws, R.M. 1979. Management of Multi Species Fisheries. Science 205, pp 267-277

APPRAISAL OF THE COMMERCIAL POTENTIAL OF
THE NEW ZEALAND DEEPWATER FISHERY

Martin S. Putterill

Associate Professor In Accountancy
University of Auckland
Auckland
New Zealand

The waters surrounding New Zealand contain some of the few remaining
underdeveloped fish stocks. As a result of 200 mile Economic Zone
legislation a relatively small country has assumed management responsibility for a significant fishery. Results from scientific
and commercial fishing activity though helping to clarify the
probable abundance of fish, has done little to inform all concerned
whether or not the EEZ declaration was a commercial bonanza. Apart
from this, there are a number of practical decisions to be made by
public sector fishery managers which affect their commercial counter-
parts, but to a degree that neither party can determine. In an attempt to resolve some of these problems, a computer model (DEFCAM)
has been developed. Consisting of 3 parts, resources, marketing and
financial, DEFCAM enables investors and managers to pretest the outcome of higher levels of investment and other problems of a practical
nature. The model maintains links between each subsystem revealing
important consequences such as the adverse impact in the form of lower
catches for all as a result of increase in effort. The paper describes
the simulation of a series of different situations. Although absolute
results are not given, there are clear indications of the factors to
which final operating results are most sensitive. In addition to these
uses, the paper briefly considers the extent to which DEFCAM may also
become useful in any partly developed fishery as an early warning
system against overfishing.

INTRODUCTION

It is easy to understand why fisheries research and management institutions
were overtaken by events when New Zealand joined the ranks of coastal states all
round the world which have declared 200 mile exclusive economic zones. The speed
and change in scale of responsibility has been exceptional. For example, sea
area up to the 800 metre depth line has increased by 250%. The number of species
with commercial potential has risen from 12 to 30 at a time when the number of
scientists engaged in offshore fishery research was increased by only 4, i.e., to
a total of 16. Even more graphic is the change in offshore catches. In 1975 in
total about 38000 tonnes was landed, a figure well below the 65000 tonnes 1979
export sales and considerably below the current estimates, approaching 500000
tonnes, of inshore and offshore potential annual catch level.

As New Zealand waters are relatively underexploited there has been pressure
exerted by a number of foreign nations to gain access to the fishery. Granting

these nations permission to fish either directly or in partnership with New
Zealand enterprises has been a major management task. Monitoring and control
duties in deeper waters have also grown considerably since the EEZ declaration.

These pressures are compounded by an awareness of how easy it is for overexploitation to occur in spite of control measures and constraints. Fortunately,
efforts are being made by the New Zealand government to determine safe limits and
to develop the necessary safeguards which are essential to prevent the decimation
of the palatable marine species of the region.

The task of a public sector fisheries manager is not simply one of protector
of species. If this were the case, conservation could be achieved by setting
ultra conservative annual catch limits. In fact, fisheries management responsibilities in New Zealand are very much wider and include:

- ensuring that the nation has an ongoing supply of fish for food consumption
  and recreational fishing
- assisting the New Zealand fishing industry to obtain optimum net earnings
  (in perpetuity)
- taking such actions as are needed to maintain or improve fishery resources
- determining the share of the total catch which other nations may take from
  New Zealand waters
- making a significant contribution towards improving New Zealand's international trading position.

From this list of responsibilities alone, it is evident that a broad knowledge
of commercial prospects for the fishery and an ability to interpret events in such
terms as their commercial impact is obligatory for the government agency. There
are other reasons why some kind of commercial appraisal facility should be available. Communication is particularly important. Government, the business community
and marine scientists have so much to gain from clearly understanding the results
of each other's actions yet traditionally have found it extremely difficult to relate to each other. Resource management decisions taken in the public sector
whether to limit catches, direct research effort, provide export incentives or
locate new fishing port facilities, all have immediate or medium term commercial
consequences.

The need to establish a sounder basis for decisions of this kind and to improve relationships between parties with interests in the fishery, gave rise to
the development of the commercial appraisal technique described below.

In 1978 a decision was made to build a model of the New Zealand deepwater
fishery incorporating commercial and biological considerations. One application
is in the field of licencing where to date no explicit attempt has been made to
gauge the impact on existing fishing companies, of the approvals given to joint
venture and foreign fleet operators. The absence of work in this field is note-

worthy in view of the Fisheries Management Division having the objective "to assist fishermen to obtain optimum net incomes from their fishing enterprises...". Internationally too, reports on commercial performance appraisal using models seem to be limited. Information was received from one government source in Australia, indicating that it was not active in any modelling involving economic considerations.[1]

Enquiries among commercial companies in New Zealand revealed interest, but no current activity in model development for fishery project investment appraisal. A paragraph from the ACMRR Working Party concerned with determining management measures[2] adds weight to the view that building a commercial appraisal model was potentially useful. The relevant section reads 'While the Working Party is aware that financial models are sometimes used to evaluate investment decisions, no specific examples were identified. Over-investment on the Mexican shrimp fishery was cited as an example of failure of investment analysis to allow for the effect of additional fishing on catch rates of fishermen'.[3]

Although appraisal of the commercial aspects of a fishery is prima facie a concern of businessmen, resource managers in the public sector are also closely connected with the process. In New Zealand where external trade relationships are particularly important, it is not too difficult to make a case for constructing a tool which would help policy makers measure the impact of competing objectives such as the trade-offs of increased fishing rights in exchange for higher import quotas for New Zealand primary produce such as beef. Clearly a single model could not be expected to deal with the whole range of concerns of such a diverse group of interested parties. Nonetheless, it seemed important to assess the information needs of each party and to see where overlap occurs, in order to define a model structure which would be useful to each separately but also help to bridge communication gaps between them. The 'philosophical' base for the model was thus to integrate and communicate important facts about the deepwater fishery in both present and future time scales.

The sections which follow are:

1. Description of the Model
2. Initial Test Results
3. Review of Potential Applications

---

*
1. Resource Management Section, Fisheries Division, Department of Primary Industry, letter dated 19 October 1979

2. ACMRR, FAO Interim Report of the Working Party on the Scientific Bases of Determining Management Measures, Rome, December 1978

3. ACMRR ibid p. 43

DESCRIPTION OF THE MODEL

The primary focus of the model is return on investment, the most common standard of measurement of business activities. Considerations such as the time value of money, distinction between fixed and variable costs and provision for incremental investment through time are some of the features which had to be incorporated. These characteristics are present in most financial investment appraisal models and are clearly described in the literature, cf. Merret and Sykes (1973), Hertz (1969), Van Horne (1972). Figure 1 below brings together these aspects of the main considerations which influence business managers when allocating risk capital.

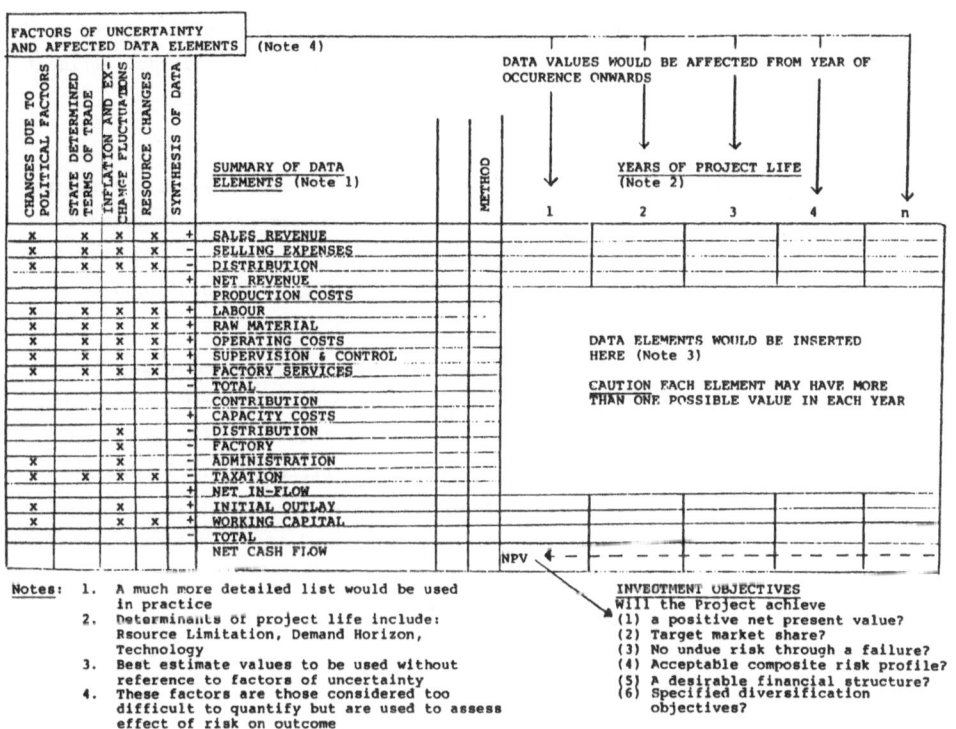

FIGURE 1. General Model of a Capital Asset Acquisition Project

The accent in Figure 1 is clearly forward looking. Apart from year 1 where actual data is used to calculate the return on an investment, the remaining years' results are estimates and as such, subject to variation. In some industries such as deepwater fishing there is a higher degree of uncertainty than in others, i.e., the probability of achieving a specified result is low. Some of the factors which give rise to greatest uncertainty are:

(a) Resource Limitation - Fishing like hunting is unpredictable - there is a limit to the availability of the raw material upon which the project depends

(b) International Markets - involvement in the sale of production to and purchase of supplies from foreign countries

(c) Government Controls - one or more of pricing, freedom to acquire raw materials, enforced quality standards, effluent and polution control requirements

Whereas these factors influence most business operations, deepwater trawling in New Zealand is particularly affected. Businessmen contemplating investment in the industry would be negligent if they failed to measure the profit and loss implications of a high incidence of one or more of these items in addition to the impact of normal trading risk.

As most fishery investment involves the acquisition of expensive capital plant with a working life of ten years or more, investment decisions have far-reaching consequences. This is true of fishing where over-capitalization has been blamed for the collapse of many fisheries. C. W. Clarke (1979) refers to nonmalleability of vessel capital as well as that of fishermen's skill, which could help to explain why unprofitable investment in fishing boats is not immediately withdrawn.

The behaviour pattern of businessmen suggests that in general they have a poor grasp of the biological aspects of fisheries. This could be the consequence of specialisation, i.e., a lack of training in the natural sciences or alternatively it is the inability of scientists to translate problems into business terms. Whoever is at fault, there has been a lack of understanding about the causes of change in catch rates as well as a regrettable tendency to budget by straight-line extrapolation. Effects on catch rates of their own increments of fishing effort, let alone those of competitors or foreign fleets, are very seldom taken into account.

These factors tend to result in optimistic estimates of future catches with obvious implications for fisheries investment decisions. To overcome this weakness and to help bring together the interests of fisheries' scientists and the business community, the model was structured to incorporate biological parameters of the target stocks as well as commercial data. A schematic outline of the components

of the commercial apparaisal model (DEFCAM - Deep Water Fishery Commercial
Appraisal Model) is provided in Figure 2.

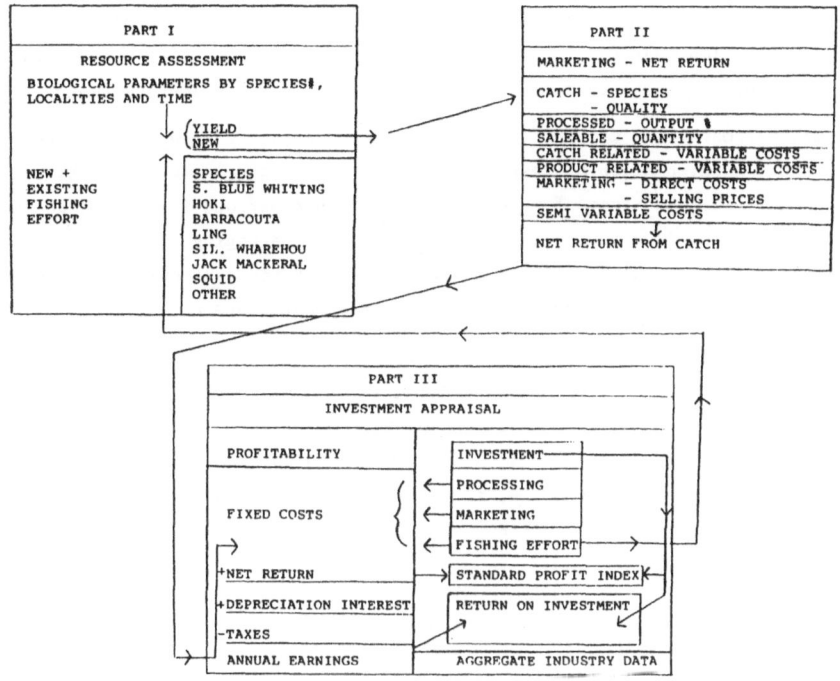

FIGURE 2. Components of the Deepwater Fishery Commercial Appraisal Model (DEFCAM)

PART I, the resource assessment module, was developed for this project by Francis and Ryan (1979). Based on the Schaefer stock production model, it has been developed in this application, to produce results consistent with the multi species nature of the New Zealand demersal fishery. Parameter estimates were derived from catch data generated by research and commercial fishing vessels since 1975. Provision was made for large and small fishing vessels, i.e., above and below 2000 G.R.T. Using an optional fishing plan in which locality and duration are specified, any fleet configuration can be used to generate the quantity of catch by species. Aggregate industry effort data is accumulated and forms part of the resource evaluation module. There is a need to check the catch estimates with current data and to revise parameter values where necessary to improve tuning.

PART II, the marketing module, converts the catch estimate into net cash revenue. In essence, a calculation of cash income is made for each 'species' of fish according to market and type of pack. Income for this purpose is defined as sales revenue per kilo minus variable costs of transport, packaging and processing. The model provides for quantity limits to be established for particular markets and for the cost and price factors to be varied from year to year.

PART III, extends the financial analysis by producing among other data, annual estimates of return on investment. Attention is also given to fixed costs, depreciation policy options to facilitate the calculation of income tax and aggregate foreign earnings both for the unit under review and for the industry as a whole. The programme, written in FORTRAN, is currently being run on a Burroughs B6700 system at the University of Auckland.

INITIAL TEST RESULTS

Obtaining appropriate test data presented a problem. The biological segment (PART I) responds to effort in increments of vessel tonnage, i.e., small vessels are <2000 tonnes and large >2000 tonnes. At the time data was needed for testing purposes, few if any vessels flying New Zealand flags were large enough to serve as prototypes even for the small category. Joint venture, charter and foreign fleet trawler sizes have been much larger, representing the view that operating in southern waters requires substantial fishing power. As the aim was to obtain test data which presented a reasonable set of relationships between fixed and variable costs, product prices and processing throughput, it was decided to use information about four 600 tonne trawlers whose catches would be processed onshore. The configuration is the size of operation which a New Zealand based company might easily contemplate. During 1979, data was obtained from the management of a locally based enterprise and coding sheets completed. It is important to note that some of the information had to be estimated as the class of vessel had at that stage not commenced fishing in New Zealand waters.

Base run output using the test data reflects an overall result over 4 years expressed in net cash flow terms. From this run, a large negative number was generated indicating on the figures used and the fishing plan selected, the operation would be unprofitable with the result that no taxes would be payable and a return on investment statistic not meaningful.

Because of data limitations, only one aspect of the problem was tested namely, variability of results. This corresponds to the concern about assessing business risk which was referred to earlier in this paper. Although data might not produce an absolutely correct result, nonetheless it contains sufficiently accurate relative information to be a useful basis for simulating variability. In all, 40 results were produced reflecting changes to one or more of the factors of cost and revenue. The pattern of changes detailed in Table A was a subjective choice covering the kind of 'what if' questions that businessmen ask. For example a typical set of questions might be - by how much would the result change if sales prices rise by x percent, costs of processing by y percent and total effort in the industry by z percent. Measurement of single factor changes help to identify the most sensitive parameters and divert attention away from less important elements. However runs which incorporate changes to more than one factor are more satisfying representations of the real commercial world.

The results of the 40 simulation runs are presented in Tables B and C. It was not easy to choose an index nor to know which graphical approach gave the clearest picture. For those businessmen used to results in "black and red" format the use of a positive index number to describe a loss situation might be difficult to swallow. Further research and discussion will help towards the selection of more appropriate ways of communicating results of this kind to a wide selection of interested parties.

Table B sets out the results of computer runs in which only one factor is changed by a stated degree. The result in dollars of net cash flow has been transformed into an index value. This is simply a 100 point scale representing the range of values generated at the third year. The choice of year 3 is arbitrary, but not unrealistic as businessmen tend to work quite comfortably in time frames of less than 5 years.

The revised single factor results are useful guides to those which have most effect on overall performance. From Table B, it is clear that selling price followed by fuel are the two factors which cause most response. Whereas selling prices fluctuate upwards and downwards, cost patterns in the industry seem destined to exert a negative influence and have been tested accordingly. The impact of change in total effort must be interpreted with care. The model produces a new catch estimate after adding the effort of the current run, to the total basic ef-

TABLE A

40 Simulation Runs - Variations Introduced in Tables B and C

Effort

EO	4 vessels are the basic effort
El	8 vessels are the basic effort
E2	16 vessels are the basic effort
E3	24 vessels are the basic effort

Selling Price

S0	basic selling prices
S1	50 above + annual linear growth of 20%-
S2	50 above + annual linear growth of 10%-
S3	50 above + annual linear growth of 15%+
S4	50 above + annual linear growth of 30%+

Production Variable Cost

VPO	basic variable cost data
VP1	VP0 + annual linear growth of 10%+
VP2	VP0 + annual linear growth of 20%+
VP3	VP0 + annual linear growth of 30%+

Distribution Variable Cost

VD0	basic variable cost data
VD2	VD0 + annual linear growth of 20%+
VD3	VD0 + annual linear growth of 30%+

Operation Variable Cost

O∅	basic operating variable cost data (wages)
O2	O∅ + annual linear growth of 20%+
03	O∅ + annual linear growth of 30%+

Fuel Cost

F0	part of seagoing fixed costs
F1	F0 + annual linear growth of 10%+
F2	F0 + annual linear growth of 20%+
F3	F0 + annual linear growth of 30%+

fort recorded for the fishery. Adding a second vessel to an unexploited fishery is a 100 percent jump in effort, but will have little effect on catch level. The converse also applies. Without defining the total of effort, the order of magnitude effort change, and the results derived therefrom cannot be useful indicators of the relative importance of this factor.

TABLE B

CHANGES	TOTAL EFFORT (E)	SELLING PRICES (S)	VARIABLE COSTS PROCESS'G (VP)	VARIABLE COSTS DISTRIB. (VD)	OPERATING (O)	FUEL (F)	INDEX RESULT (Year 3)	Δ % INDEX BASIC
1.	x							0
2.		x						-3
3.			x					-6
4.		x						-54
5.		x						-27
6.			x					+37
7.			x					+59
8.			x					-8
9.			x					-16
10.				x				-24
11.				x				-10
12.				x				-19
13.					x			-5
14.					x			-6
15.					x			-11
16.						x		-24
17.						x		-35
BASIC	x							0

Basic

Most businessmen, if pressed, would be able to indicate the direction of change over the short term to more than one cost or revenue factor. Examples of such combinations produce a range of results whose magnitude and variation is described in Table C.

In both Tables B and C, the dotted line marked 'Basic' signifies the result using current best estimates on a single figure basis. As mentioned earlier, the "basic" result was a negative number of no small order. The pattern of cash flows revealed from this limited analysis has facets of interest to resource managers and investors. Some of these considerations are examined in the next section.

TABLE C

SENSITIVITY ANALYSIS OF FACTOR COMBINATIONS

INDEX RESULT AT YEAR 3

Basic

REVIEW OF USES TO WHICH THE <u>DEFCAM</u> MODEL MIGHT BE PUT

The acid test of a new model is its ability to foster a better quality of decision making than is being achieved with existing methods. Until some practical
experience is obtained, justification for development must rely on normative
grounds. The parties of potential users include the following:

- corporate management in private sector organisations who are or might
  become active in fishing, processing, marketing or financing that sector
- public sector administrators such as the Fisheries Management Division,
  Fisheries Research Division, the Department of Trade and Industry and
  the Treasury
- quasi public bodies like the Fishing Industry Board

- statutory organisations such as the Department of Scientific and Industrial Research (DSIR) and National Research Advisory Council (NRAC)

Examples of the decisions confronting each group include:

- in the case of corporate management, new investment in or disinvestment from fishing related projects and the maintenance of optimum results from resources committed to the sector
- public sector administrators who are responsible for orderly development of the marine resources of the zone through the licensing of foreign fleets, approval of joint ventures and advice given to government on legislation and regulation based on research and management experience
- the Fishing Industry Board which must decide the development directions most beneficial to the industry in such fields as new markets, produce range, technological innovation
- the DSIR and NRAC defining and supporting research in many aspects of the fishery and who must be aware of areas of real and probable return from research investment.

As the actions of one group tend to affect others, a way of ensuring that the consequences are measured and communicated is an obvious need. For public administrators and businessmen alike, the fishery related decisions require that a constant watch be kept to 'determine whether improvements introduced in one sector properly compensate for deficiencies brought about in other sectors of the system', c.f. Churchman (1968). This is particularly true in the case of conservation control procedures. It has been shown that it is the aggregate of fishing effort plus new effort which produces levels of catch. Unless those who approve applications by foreign fishing interests are informed about the economics of the fishery they may take actions which do significant harm to an existing party such as a local fishing company. In New Zealand it is cause for alarm that the level of foreign fishing activity in its southern waters has been allowed to rise to the same intensity as it was when open access was halted a few years ago, because of fear of overfishing.

There is no doubt that major national considerations impinge on every fishery. Access to overseas markets for agricultural produce, foreign exchange earnings from exports and import replacement are some of the issues of this kind. Fiscal regulation, export incentives and tariffs play a part in fostering national interests. Introduction of fiscal programmes without knowing how much support is actually needed is both costly and misdirected.

Last in this list of management issues, but by no means least important is
the question of how much fish should be caught. Clark (1979) has given the topic
a very thorough review, but the problem of practical application still remains.
If the aim of fishery policy is to allow an economic return, how is the yield to
be assessed when nations have markedly different norms and objectives? Pressures
of local and international politics are likely to disturb most carefully calculated economic or biological based regimes. Administrators are likely to need
more urgently than ever some benchmark against which to judge the degree to which
general political considerations are making inroads into sound conservation
principles.

These decisions are only part of the web of concerns which make the fishing
industry challenging and frustrating. In this regard Forrester's comments on
complexity are worth stating, 'Complex systems are counterintuitive. They respond
to policy changes in directions opposite to what most people expect. We develop
experience and intuition almost entirely from contact with simple systems, where
cause and effect are closely related in space and time. Complex systems behave
very differently, c.f. Forrester (1969).' All concerned need to look at the
DEFCAM model to see whether it will help in the communication process, in direct-
ing attention, in good time, to problem areas and helping channel efforts in a
positive way. The three 'ecosystems', marine resource, commercial and political,
need a key to relate one to the other. The commercial appraisal model has the
latent capacity to meet some of these needs. Businessmen may use it to judge the
potential investment returns, fisheries management to monitor the commercial performance of New Zealand fishing interests. Although the marine biologists may
disagree, using the commercial performance of the coastal state as a barometer
may be as effective a measure of the condition of the stock as any more scientific
measures of abundance and well being.

This is particularly true in the New Zealand situation where return on investment objectives are more likely to be conservative and thus dependent on higher
catch rates per unit of effort than other countries in which capitalist principles
are of less importance. Although one is hesitant to add yet another category of
fishery management regime, it does seem that the Host Nation Commercially Sustainable Yield (HNCSY) could be an important measure in developing fisheries such as
New Zealand. The DEFCAM model could prove useful as a means of monitoring the
progress of a fishery for which a commercial measurement criterion is chosen.

CONCLUSIONS

In this rapid sketch of fishery management decision responsibilities, it has not been possible to deal with all aspects of this complex task. The aim was to show that there is common ground in the content and consequences of decisions taken by public and private sector managers. Furthermore, that a commercial appraisal model of the kind that has been demonstrated, offers the prospect of more holistic and realistic decision-making and better communication. It may not be overly elegant but, because of relevance, it stands a chance of harnessing a wide panel of users.

ACKNOWLEDGEMENTS

During the past year a working party of the National Research Advisory Council with Professor E. C. Young as chairman has been considering the consequence for fisheries research and management of the 200 mile Exclusive Economic Zone declaration by New Zealand. As a member of that committee, it has been possible to identify many issues and problems among which the absence of commercial appraisal was a singular deficiency and prompted the work.

Although responsible for defining the approach structure and end uses, I would not have been able to carry out the project without the help of Dr. Bob Francis and Dr. Paul O'Connor - the former by developing the biological simulation (Part 1) and the latter by carrying out the bulk of responsibility for programming. This co-operation was very pleasant, constructive and a good illustration of the benefits of interdisciplinary approaches to problems of this kind. John Paynter helped the project in its early stages, Jill Reid did the typing at the usual eleventh hour and Duncan Waugh, Director of Fisheries Research Division, generously met the out-of-pocket costs.

REFERENCES

Clark, C.W.   1979.   Towards a Predictive Model for the Economic Regulation of
Commercial Fisheries, Univ. of British Columbia, Department of Economics,
Resources Paper #40.

Churchman, C.W.   1978.   Challenge to Reason, McGraw-Hill, N.Y., p 7

Forrester, J.W.   1969.   Overlooked reasons for our social troubles, Fortune,
Dec., 191-192

Hertz, D.B.   1969.   New power for management, McGraw-Hill, N.Y.

Merrett, A.J. and Sykes, A.   1973.   The finance and Analysis of Capital Projects,
Sec. Edition, Longman Group, London

Ryan, C.M. and Francis, R.C.   1979.   A simulation model of New Zealand's deep water
demersal fishery, Fish. Res. Div., Min. Agricult. and Fish., Wellington, N.Z.

Van Horne, J.C.   1972.   Financial Management and Policy, Sec. Ed., Prentice Hall

SUBINJURIOUS MAINTENANCE STRATEGIES FOR
OPTIMAL CONTROL OF NEMATODES

Hugh N. Comins
Environmental Biology Department
Research School of Biological Sciences, A.N.U.
Canberra, A.C.T. 2600

Brian R. Trenbath
Centre for Environmental Technology
Imperial College
London SW7
England

We consider pest species with a threshold form of density-dependence for intra-specific competition and crop damage. Under certain circumstances the optimal control of such pests requires that their numbers be kept below both crop-injury and competition thresholds, although eradication is impossible. We derive optimal long-term control strategies for this particularly simple case. The technique is applied to a model of the root-knot nematode <u>Meloidogyne javanica</u> in resistant-susceptible crop mixtures. The results emphasise the importance of understanding the dynamics of root infection.

INTRODUCTION

In many pest-control situations the potential damage by pests far outweighs the cost of control. When this characteristic is combined with certain others, including a high threshold for pest intraspecific competition, it becomes very likely that the optimal strategy will be one of sub-injurious maintenance (henceforth SIM). In a SIM strategy the pest density is never allowed to exceed the threshold for significant crop damage, although no attempt is made at eradication. A mathematical model which is restricted to SIM strategies will clearly be unusually simple, since it will not require the use of crop damage functions and may also be able to ignore pest intra-specific competition. It is worthwhile to consider such models, since the low requirement for data collection may make it practical to employ them where a complete system model is not economically feasible.

In this paper we derive long-term strategies for the class of SIM models in which the pest population can be represented by a single number at some time of year, for example, the number of overwintering eggs. There are no intrinsic limitations on the number of control options used, and these can be either discrete or part of a continuous range. The techniques described can be used to determine control strategies if the basic SIM conditions have been shown to apply, or if this is not certain they can provide intuitive explanations for numerical results which may arise from a full crop-pest population model.

These concepts are illustrated in a model of root-parasitic nematodes interacting with a mixed crop of susceptible and resistant plants, in which a SIM strategy was found to be optimal empirically. In this model the general SIM techniques predict under what circumstances one should use mixtures, as opposed to pure susceptible or resistant crops. The results depend on the strength of "crypsis" effects, in which resistant components of a mixture screen susceptible plants from nematode attack by absorbing a proportion of the dispersing larvae. Naturally such effects tend to shift the optimum strategy towards employing mixtures. Models of crypsis have a general significance in biological control since we would be led to similar mathematical descriptions for a number of functionally similar mechanisms, as for example the protection of carrots from carrot-fly by intercropped onions.

NEMATODE MODEL

The root-knot nematode <u>Meloidogne javanica</u> has a wide range of agronomically important host species, including various vegetables and tobacco (Taylor 1976). In areas where a dry season unsuitable for plant growth alternates with a wet cropping season the nematode survives the dry season as resistant eggs in the soil. Most of these hatch near the beginning of the wet season, and the resulting second-stage larvae disperse in search of plant roots. Sexual development is environmentally determined, nearly all individuals being female at low density. Females which find susceptible plants may eventually produce new egg-sacs. Some of these lie dormant, while the eggs in others hatch to produce a new generation of dispersing larvae. Approximately three such generations occur during the growing season of an annual crop.

When susceptible and nematode-resistant plants are intercropped it is found that any yield loss in the former is to some extent compensated by enhanced growth of the resistant plants, because of reduced competition for resources. This process is mimicked by a plant competition model of the type proposed by de Wit (1960), the yield per unit area of component A in a mixture of A and B being given by

$$y_{AB} = \frac{k_{AB}p_A}{k_{AB}p_A + p_B} \, y_{AA} \tag{1}$$

where y_{AA} is the monoculture yield per unit area of A, p_A is the proportion of A, $p_B = 1-p_A$, and k_{AB} is a measure of the competitive advantage of A over B. When p_A is small this model implies that "A" individuals grow k_{AB} times as large as in a monoculture of the same density; on the other hand, y_{AB} becomes equal to y_{AA} for $p_A = 1$. For plants competing for the same resource it is assumed that a similar equation holds for component B, but that $k_{BA} = k_{AB}^{-1}$.

In the model developed by Trenbath (1977) (see Fig. 2 of reference) the mono-culture yield (y_{AA} if A = susceptible, B = resistant) and competition coefficient $k_{AB} = k_{BA}^{-1}$ are assumed to decrease when nematodes are sufficiently numerous [for the decreasing functions used, see Trenbath (1977) and Comins & Trenbath (to be published)]. Except for the considerations of the next section the model does not describe the multiple generations of the nematode, simply ascribing a certain degree of yield loss and retardation of root growth (i.e., reduction in k_{AB}) to a given initial number of nematodes. A similar lumping procedure applies to nematode intra-specific competition effects, which are also described by a decreasing sigmoid-shaped function. This latter function incorporates a dependence on the average size of susceptible host plants, so that smaller plants give a lower maximum multi-plication rate of the nematodes and a lower threshold density for the onset of intraspecific competition.

DYNAMICS OF INFECTION

A simple view of the infection process suggests that the number of nematodes finding susceptible hosts should be proportional to the fraction p of susceptible plants in the crop. However, there are a number of mechanisms which can produce a more complicated dependence on p. These are discussed in detail here because of the important effect they have on the selection of SIM strategies.

The first mechanism is suggested by an experiment of Sibma et al. (1964), as discussed by Trenbath (1977). In these results resistant barley plants provide a disproportionate amount of protection from Heterodera avenae to intercropped sus-ceptible oats, apparently due to their earlier root development. We call such pro-tection of intercropped susceptible plants "crypsis" and define the "crypsis fac-tor" f to be the ratio of the actual infection rate to that expected from simple dilution. In this case (assuming random search by dispersing larvae) we have

$$f = [\frac{p}{p+g(1-p)}]/p = [p+g(1-p)]^{-1} \tag{2}$$

where p is the proportion of susceptible plants and g is the relative searching ef-ficiency for resistant plants compared with susceptible plants. In particular, if $g = 1$ the two types of plants are equally easy to encounter and we have $f = 1$ (no crypsis).

This model assumes that the larvae have sufficient reserves so that they all encounter a plant, and that those encountering resistant plants penetrate them rath-er than continuing their search.

A second possibility is that nematodes are attracted by substances exuded by the plant roots and that these substances vary in quantity or composition between the two types of plant. This process can also be described by Equation 2. Note that both processes can operate in reverse giving $f > 1$, with infection rate exceeding that expected from simple dilution.

It can also be shown (Comins and Trenbath, to be published) that purely random infection can result in a crypsis effect if repeated for two or three generations of nematodes during a single cropping season. The resulting crypsis factor is always less than one (i.e., positive crypsis).

It is obviously necessary that the crypsis factor should be unity in a pure susceptible crop, since there can be no crypsis without resistant plants. Both the above crypsis functions have this value when $p = 1$. Also for positive crypsis they both decline monotonically to a value between 0 and 1 as the proportion of susceptibles tends to zero. This behaviour concurs with that of the empirically determined crypsis factor used by Trenbath (1977), which has a linear dependence on p:

$$f = s + (1 - s)p \tag{3}$$

where s = minimum relative attack on susceptible plants (compared with monoculture). Both Equations (2) and (3) are used later, in order to investigate the effect of a curved crypsis function as compared with a linear one. It will be shown that for a function with a particular value of f at $p = 0$, an upwards curving form such as Equation (2) has a much stronger effect in promoting the use of mixed crops.

Since the crypsis factor multiplies the nematode infection rate it appears in three places in the model equations (Comins & Trenbath, to be published). It multiplies the nematode reproduction rate in the crop, and the number of nematodes in the sigmoid functions for yield loss and root growth retardation in susceptible plants.

LONG-TERM STRATEGIES

In developing long-term strategies for the mixed-crop nematode system we make the assumption that all resistant eggs are indistinguishable, so that all eggs present at the end of the dry season have the same likelihood of hatching. In practice this assumption is not critical since relatively few eggs remain dormant and viable for more than one dry season. The model equations can now be written in the following form

$$N_{t+1} = R(N_t, p)$$

$$Y_t = T(N_t, p) \tag{4}$$

where $\underline{N}_t$ is the number of resistant eggs just before the start of the $\underline{t}$th cropping
season, $\underline{Y}_t$ is the total yield for the $\underline{t}$th year, $\underline{p}$ is the proportion of susceptible
plants grown and R ($\cdot$) and T ($\cdot$) are specified functions. Thus the proportion of
susceptible plants combines with the number of resistant eggs to determine both the
total yield for the year and the number of resistant eggs in the next year.

Suppose that the objective of long-term management is to maximise total profit
over a certain number of years (say $\underline{m}$ years). The optimal crop mix for each year
can be determined as a function of the number of resistant eggs present using the
technique of dynamic programming (Bellman 1957). The optimal strategy is first
determined for the $\underline{m}$th year. Using this result the strategy for the (m-1)th year
can be determined, and the process may be repeated to give strategies for each of
the $\underline{m}$ years. A fundamental result of dynamic programming theory (idem., p. 238)
states that the year one strategies in this process converge to a stable "long-
term" strategy as $\underline{m}$ tends to infinity. Since this convergence occurs rapidly in
the nematode model the long-term strategy closely approximates the optimal strategy
for any reasonable time-horizon, and we will therefore take this limit rather than
considering a specific value of $\underline{m}$.

In the initial calculation of optimal long-term strategies it was found that
an eradication strategy predominated. This required that pure resistant crops be
grown until the nematodes were exterminated, whereupon the susceptible crop could
be grown at full yield. Since eradication would not work in practice this aberrant
strategy is excluded by introducing a small amount of nematode immigration. This
causes a qualitative change in the strategy which is effectively independent of the
actual value of the immigration rate. We therefore assume a standard value of one
nematode per plant per year.

Some dynamic programming results are shown in Figure 1. The graphs specify
the proportion $\underline{p}$ of susceptible plants to be grown for each initial number of nematodes (as determined by various monitoring techniques). Once the system has set-
tled down much of this information is of only academic interest, since the control
strategy keeps the nematode population within narrow limits. However, should a
large perturbation occur, the graph gives the appropriate crop mixture to use
(provided such perturbations are very infrequent, otherwise the dynamic programming method must be modified). Simulation shows that in the absence of perturbations the strategy of Figure 1a (a case with no crypsis) is an alternation between
pure resistant and pure susceptible crops. Figure 1b in which resistant roots are
assumed to be five times as attractive (cf. Equation 2) gives an alternation between a pure resistant crop and a mixed crop with 78% susceptible plants. Thus a
positive crypsis effect can cause a mixed crop to be used instead of a pure resistant one. This result is predicted by the SIM analysis in the next section, in
which we elucidate the dependence of the optimal strategy on the critical model

parameters, including crypsis.

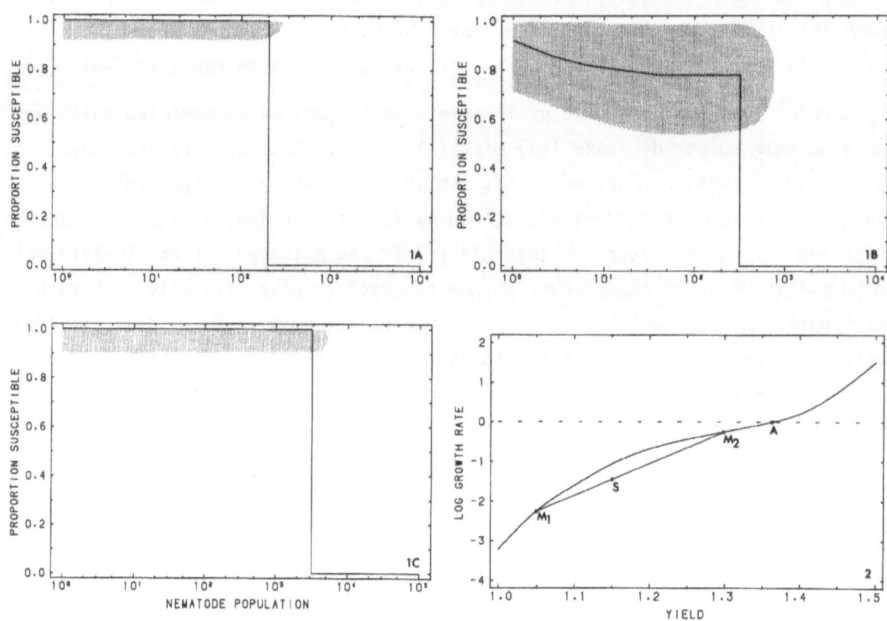

FIGURE 1. Dynamic programming optimal strategies for the nematode model [E = 0.5, M = 0.9, D = 0.8, m = 10, A_R = 2, A_Y = 1.5 (see Equation 10); competition, yield and root damage functions have 1% point at population density 10^5 per plant, and are appropriate to a susceptible but intolerant host (see Comins and Trenbath, to be published)]: (a) no crypsis, (b) g = 5 in Equation 2, (c) tolerant host, no crypsis. Shaded areas show strategies with long-term yield not more than 0.02 yield units [i.e., 2% of pure resistant crop yield in one season] short of optimum.

FIGURE 2. Graph of logarithmic pest growth rate versus yield, illustrating construction for obtaining the results of a 2-mixture strategy.

SUBINJURIOUS MAINTENANCE

If the local population of a species is to remain bounded its logarithmic growth rate must average to one in the long term. This can be achieved through one or more of the following mechanisms: immigration, density-dependent reproduction and survival, and (in the case of a pest species) density-dependent control. In this analysis we postulate that there is a range of pest population in which only the last mechanism is significant, and that it is advantageous to keep the pest population within this range (as is done in the strategies of Figures 1a and 1b).

The existence of a range of endogenous density-independence is assured if the reproduction and survival rates have a threshold form of density dependence (e.g., decreasing sigmoid or Hassell (1975)) and if the migration rate is sufficiently small. It is clearly advantageous in terms of control effort to keep the pest population much larger than the immigration rate, since immigration produces a high effective growth rate at low densities. On the other hand, if the potential pest damage is much greater than the cost of control, then the population should be kept below the level causing significant crop damage.

In cases where pest competition acts through the depletion of the harvestable part of a plant it is clear that the requirement of negligible damage precludes the use of density-dependent reproduction and mortality for control purposes, so that it is indeed necessary to keep the pest population within the endogenous density-independent range. However, where a pest acts on some other plant part, such as the roots, it is possible to observe tolerance, in which host plants are not seriously affected by the equilibrium density of pests (Nusbaum et al., 1971). It may then be desirable to keep the pest population slightly above the crop injury threshold, trading a small loss in yield for a large reduction in pest population growth rate. An example of such a "slightly injurious maintenance" strategy is illustrated in Figure 1c. Note that the threshold population for growing a resistant crop is much higher in this case than in Figure 1a. This leads to yield losses of up to 2.8%, which in this case corresponds to a 27% reduction in nematode growth rate. The analysis of "slightly injurious maintenance" is more complex than that of sub-injurious maintenance, so we will leave consideration of tolerant host plants until later. For the moment we concern ourselves only with intolerant hosts, which in any case suffer the most economically important pest problems. We may thus assume that it is advantageous to keep the pest population within the range in which control measures are the only form of density-dependence and crop damage is negligible. Within this range Equation 4 can be simplified to give

$$N_{t+1} = N_t \exp (L(p))$$

$$(5)$$

$$Y_t = T_o(p)$$

where $\underline{N}_t$ is the resistant egg population at the start of growing season $\underline{t}$, $\underline{Y}_t$ is the total yield, $\underline{p}$ is the proportion of susceptible plants, $L(\cdot)$ expresses the logarithmic growth rate of the nematodes as a function of $\underline{p}$, and $T_o (\cdot)$ gives the total yield from crop mix $\underline{p}$ in the absence of nematodes.

The optimal long-term SIM strategy is required to maximise the average yield $\underline{Y}_t$ while ensuring that the nematode population shows no long-term increase. For the moment we will ignore the limits to the validity of Equation (5), so that there are no restrictions on the short-term variation of $\underline{N}_t$. The most general strategy then consists of a set of crop mixes p_1, p_2, ..., p_k which are used with frequencies ϕ_1, ϕ_2, ..., ϕ_k, and it is required to maximise

$$\bar{Y} = \sum_{i=1}^{k} \phi_i T_o(p_i) \qquad (6)$$

subject to

$$\bar{L} = \sum_{i=1}^{k} \phi_i L(p_i) = 0 \qquad (7)$$

It can readily be shown (Comins and Trenbath, to be published) that the optimal strategy is either a single crop mix p^* such that $L(p^*) = 0$, or else an alternation of two crop mixes $\underline{p}_1$ and $\underline{p}_2$. This result allows the optimal SIM strategy to be derived by a simple graphical construction. First of all it is necessary to combine Equations (5) to give the logarithmic growth rate $\underline{L}$ as a function of total yield $\underline{Y}$, as illustrated in Figure 2. This graph expresses the result of any single-mix strategy, and the yield for the optimal strategy of this type is given by the intersection of this curve with the line $L = 0$.

Now consider any two points M_1 and M_2 on the L(Y) curve. It is an elementary result of coordinate geometry that any point S on the straight line joining these two points has $\underline{L}$ and $\underline{Y}$ coordinates

$$L_s = \phi L_1 + (1 - \phi)L_2$$

$$Y_s = \phi Y_1 + (1 - \phi)Y_2 \qquad (8)$$

where M_1 has coordinates (Y_1, L_1), M_2 has coordinates (Y_2, L_2) and ϕ is a number between 0 and 1. By comparing this result with Equations (6) and (7) we see that every alternating strategy which uses the two mixes corresponding to M_1 and M_2 can be represented as a point on the straight line M_1M_2 .

Hence all possible one and two mix SIM strategies may be represented either by a point on the curve L(Y) or else a point on a straight line joining two points on this curve. The optimal SIM strategy is found by taking the rightmost (highest yielding) strategy which lies on the line $L = 0$. The extension to points on straight lines will clearly greatly enlarge the set of possible average results,

however there are still very definite limitations. A little experimentation will
show that the optimal strategy must either lie on the original curve or on one of
the four types of tangential lines shown in Figure 3. This observation forms the
basis for the following classification of optimal SIM strategies.

TABLE 1

TYPE	IF $N < N_o$	IF $N > N_o$	FIGURE
EQUIL	TARGET N_o	TARGET N_o	
MIN-MAX	PURE SUSCEPTIBLE	PURE RESISTANT	3(a)
MIN-MIX	MIXED CROP	PURE RESISTANT	3(b)
MIX-MAX	PURE SUSCEPTIBLE	MIXED CROP	3(c)
MIX-MIX	MORE SUSCEPTIBLE MIXED CROP	MORE RESISTANT MIXED CROP	3(d)

The table also shows a method of implementing each strategy under conditions
where Equations (5) are subject to random variations. The one-mix "equilibrium"
strategy can be achieved by picking a population size N_o to be the desired equilibrium and always choosing p so that the predicted N_{t+1} is equal to this target
value. The two-mix strategies on the other hand have an obvious interpretation in
terms of a threshold population size N_o, the more resistant mix (or resistant pure
crop) being used whenever the population exceeds N_o. These interpretations minimise the variation in nematode numbers produced by a given SIM strategy; the value
N_o being chosen so that the extreme population sizes are well away from the immigration rate and crop injury threshold. Otherwise N_o is chosen to be as large
as possible to save on any initial control effort. It may not be possible to
choose an N_o if control actions can change the population by an amount comparable
to the density independent range. In such cases the present analysis is not applicable.

Finally we consider some generalisations of the graphical technique. In general there may be control costs independent of yield loss. Thus yield in Figure 3
should be replaced in general by total profit P.

Figure 4 shows how to deal with different types of control variables. In Figure 4a the control variable is partly discrete and partly continuous. Figure 4b shows the construction for multiple control variables, that is

$$N_{t+1} = N_t \exp (L(c_1, c_2, \ldots))$$
$$P_t = T_o (c_1, c_2 \ldots) \tag{9}$$

where $\underline{c}_1$, c_2 etc. are continuous or discrete control variables. An example in the nematode case is the use (or not) of nematicide at the time of sowing. Note that adding more control variables does not make the technique any more complicated; however a pest population which had two or more distinct components year round would do this since Figure 4 would then need to be plotted in three or more dimensions.

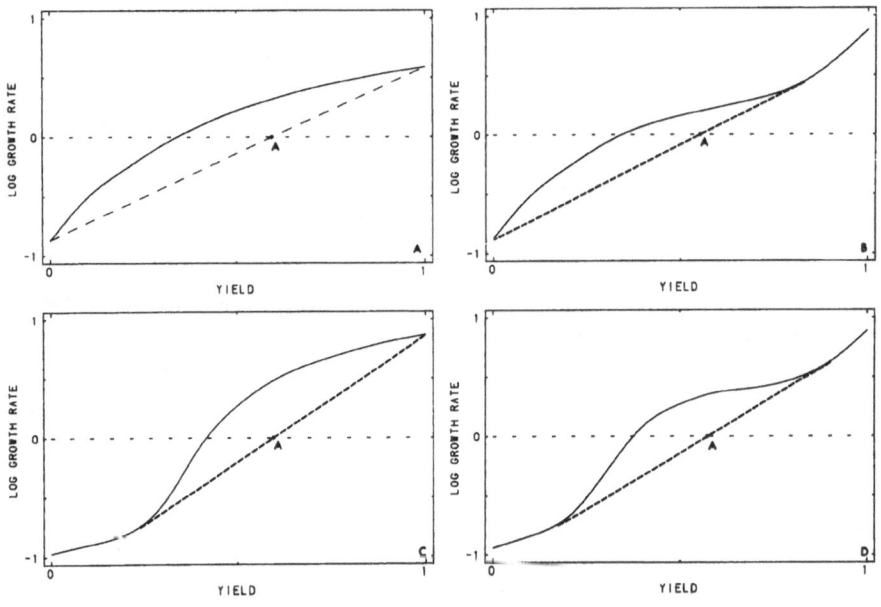

FIGURE 3. Types of optimal 2-mixture strategies (see TABLE 1).

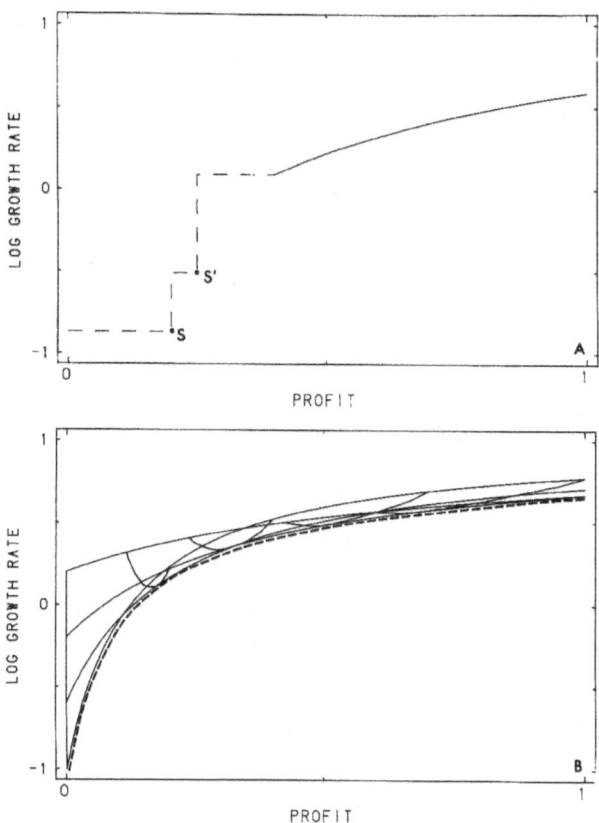

FIGURE 4. Generalisation of Figure 2 to (a) a partly continuous and partly discrete control variable, (b) multiple control variables (in this case 2).

GENERAL NEMATODE MODEL

Figure 1 shows that under certain circumstances crypsis effects can lead to mixed crops being part of the optimal control strategy. In this section we give a general analysis of the effects of crypsis using the SIM techniques just described. From Trenbath's (1977) model we have the following logarithmic nematode growth and total yield

$$L(p) = \ln \left[EHm \cdot \frac{pA_R}{pA_R + (1-p)} f + DE(1-H) \right]$$

(10)

$$Y(p) = \frac{A_Y A_R p + (1-p)}{A_R p + (1-p)}$$

where p = proportion susceptible

E = egg survival rate

H = hatch rate

D = dormancy survival rate

m = nematode intrinsic growth rate

A_R = relative growth of susceptible roots (nematode free)

A_Y = relative monoculture yield of susceptible plants (nematode free)

f = crypsis factor (see Equations 2 and 3).

These equations can be simplified by scaling and lumping parameters. First we define the scaled yield w as follows:

$$w = (Y-1)/(A_Y-1) = A_R p/(A_R p + (1-p)) \qquad (11)$$

Since this equation is a linear function of Y it is possible to substitute w for Y in Equation 9, and so the SIM analysis of the last section can be performed equally well using the scaled yield. Substituting in Equation (10) we get

$$L(w) = \ln (\mu wf + 1) + \ln M_o \qquad (12)$$

where M_o = DE(1-H) is the nematode rate of growth (less than one) in a pure resistant crop and μ = mH/(D(1-H)) is the ratio minus one of the growth rates in pure susceptible and pure resistant crops. Since the second term $\ln(M_o)$ is a constant we find it more convenient to plot

$$U(w) = \ln (\mu wf + 1) \qquad (13)$$

and then to find the intercept with U = $-\ln M_o$. Thus varying M_o allows us to vary the horizontal line in the analogue of Figure 2.

In the case without crypsis (f=1) the functional form of U(w) is particularly simple (see Figure 5a), being simply a shifted log function. Figure 5a also shows the straight line on which the optimal strategy must always lie, regardless of the value of M_o. This is necessarily a "min-max" strategy in the classification of Table 1, and uses only pure resistant and pure susceptible crops. This result agrees with Figure 1a.

Inverting Equation (11) to find p as a function of w, and substituting this result in Equation (3) gives the following expression for linear crypsis:

$$f = 1 - (1-s)(1-w)/(1-w + w/A_R) \tag{14}$$

where s = minimum relative attack on susceptible plants. The curve U(w) is now a function of three parameters: μ, A_R and s. As an example of this we plot (in Figure 5b) U(w) for the case μ = 112.5, A_R = 2, s = 0.2. Clearly the change produced by crypsis is insufficient to make the "min-max" strategy non-optimal. Figure 5c shows on the other hand that using the upwards-curving crypsis function of Equation 2 (with the same crypsis effect as p → 0) gives either a min-mix or equilibrium optimal strategy (the former being illustrated in Figure 1b). The appropriate expression for this case is obtained by inverting Equation (11) to find p and substituting in Equation (2):

$$f = 1 - (1-g^{-1})(1-w)/(1-w + w/(gA_R)) \tag{15}$$

Defining $s = g^{-1}$ to be the minimum relative attack on susceptible plants (as in Equation 14)

$$f = 1 - (1-s)(1-w)/(1-w + w/A_R') \tag{16}$$

where

$$A_R' = A_R/s$$

By pure coincidence Equation (16) is identical in form to Equation (14) except that the relative root growth of susceptibles must be multiplied by s^{-1}. This enables us to illustrate the strategic consequences of both crypsis expressions in the same figure (Figure 6).

This figure summarises the geometric properties of the L(w) curve for various combinations of the nematode growth rate μ and the root growth ratio A_R (for derivation see Comins & Trenbath, to be published). The possible strategies are obtained by comparison with Figure 3, and in some cases depend on the value of M_o. Although the equation for inflection points admits of up to four inflections, the most observed is two (within the crescent-shaped area in Figure 6) and there is generally at most one inflection. Note that the value of A_R is quite critical in determining strategies, so that the upwards-curving crypsis function of Figure 5c, which effectively has A_R = 10 (cf. Equation 16) has a much greater effect than the linear crypsis of Figure 5b, which only has A_R = 2. For general curved crypsis functions an indication of the effect of curvature is provided by dividing A_R by the ratio of the slope of the curve at p = 0 to the slope of a linear crypsis function (Equation 6) having the same value at p = 0.

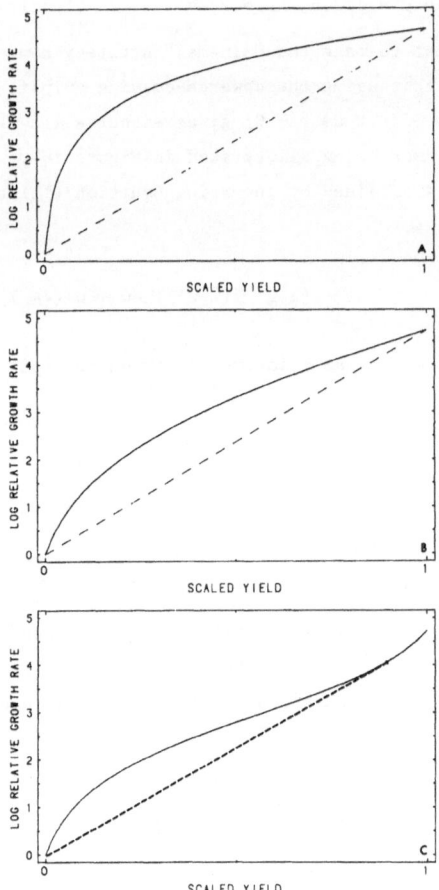

FIGURE 5. Logarithmic growth rate versus scaled yield for the generalised nematode model (Equation 10): (a) no crypsis, (b) linear crypsis, s = 0.2 in Equation 3, (c) g = 5 in Equation 3 .

Some additional attributes of the model for the parameter combinations of Figure 6 are illustrated in Figure 7. Figures 7a and 7b show the scaled yields w̲ for each component of the mixture strategies (equal for equilibrium strategy) and demonstrate the generally smooth gradation between different regions of Figure 6.

Figure 7c shows the difference in scaled yield $\underline{w}$ between optimal "min-mix," "mix-max" and "mix-mix" strategies and the best of "min-max" and "equilibrium." A median value of $\underline{M}_o$ is used (equal to $(\mu+1)^{1/2}$). The small differences observed here document the fact that it is difficult in this model (with reasonable $\underline{A}_R$) to construct curves like Figures 3b–3d which depart significantly from linearity. That is, if crypsis is strong enough to favour mixed cropping at all then it is satisfactory to use a mixed crop alone (i.e., the equilibrium strategy). Thus we conclude that in the present model there is a single important choice, between a min-max strategy when crypsis effects are small, and an equilibrium strategy when they are large. At the dividing line between these options (joining the peaks in Figue 7c) the choice of strategy is unimportant, even though strategies of types 3-5 may be marginally optimal.

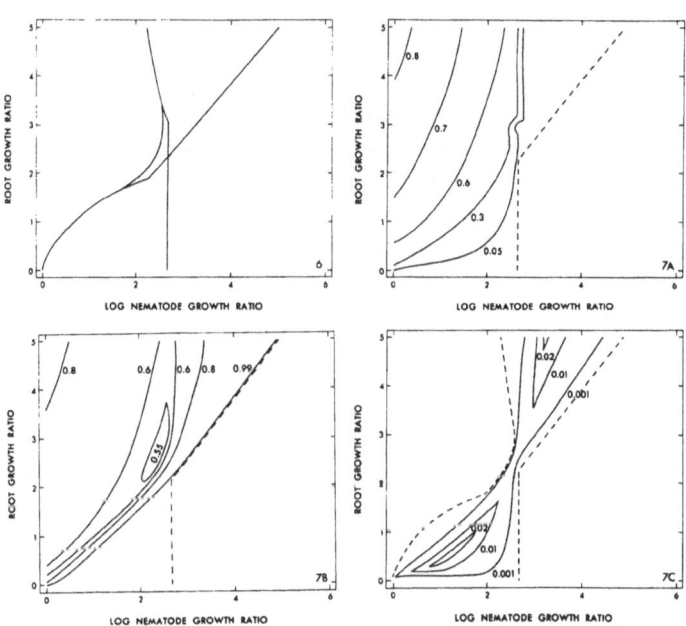

FIGURE 6. Type of optimal strategy as a function of root growth ratio A_R, and nematode growth ratio μ (scale is $\ln \mu$). Central crescent: equil or mix-mix; right side: min-max; bottom left: equil or mix-max; left side: equil; top centre: equil or min-mix.

FIGURE 7. Generalised nematode model results as a function of root growth ratio A_R and nematode growth ratio μ : (a) scaled yield of more resistant crop (see Equation 11), (b) scaled yield of more susceptible crop, (c) difference between optimum average yield and best of "equil" and "min-mix" average yields (all scaled).

Dynamic programming results for intolerant crop plants (as in Figures 1a and 1b) are basically in agreement with Figure 6, although numerical inaccuracies cause some variations in the central region. These variations are to be expected since, as stated above, the decisions involved are very marginal.

For intolerant crop plants (as per Figure 1c) the decision boundaries must be considerably revised. It was found empirically that laissez-faire strategies (growing pure susceptible all the time) displace equilibrium and mix-max strategies in the left part of Figure 6, equilibrium strategies displace min-mix and mix-mix displace min-max. If these effects are large the SIM analysis will be invalid and it will be necessary to resort to numerical methods such as dynamic programming. Fortunately these exceptional cases are likely to be the least severe pest problems, since they involve highly tolerant crop plants.

CONCLUSION

It is possible to predict from certain qualitative features of a crop-pest system that the optimum long-term control strategy is one of sub-injurious maintenance (SIM) in which the pest population is always kept below the crop-injury threshold, but eradication is not attempted. With this pre-condition the calculation of the optimal strategy is greatly simplified, and may be performed graphically.

The SIM analysis assumes that the optimal control strategy operates within a range in which pest reproduction and survival are density independent. Thus we require that the immigration rate is low, that intraspecific competition has a threshold dependence on pest density, that crop damage can greatly exceed the cost of control, and that the host plant is intolerant of the pest (i.e., is significantly damaged by an uncontrolled pest population). The last two conditions simplify the calculations by precluding the use of pest intra-specific competition for control.

Within the sub-injurious density-independent range both the pest population growth rate and crop yield are independent of pest numbers. Thus the optimal control strategy can be determined from a sub-model of the original system, without detailed evaluation of pest competition effects, crop damage functions or immigration rates. A further condition for the use of this analysis is that the resulting optimal strategy should operate completely within the density-independent range. This may not be satisfied if very effective control options are allowed (e.g., pesticide application when there are inadequate refuges).

We apply the SIM analysis to Trenbath's (1977) model of a nematode population interacting with mixed crops of resistant and susceptible plants. It is found that there is a single major choice to be made; between an "equilibrium" strategy, where essentially the same mixed crop is used each year, keeping the nematode population almost constant, and a "min-max" strategy, where either a pure resistant or a pure susceptible crop is grown, depending whether the nematode population is above or below a certain threshold. The equilibrium strategy may be preferred when the resistant component of a mixture reduces nematode attack on the susceptible plants more than would be expected from simple dilution; an effect we call "crypsis."

Control using resistant plant varieties carries with it the risk of selecting "resistance breaking" genes in the pest. It can be shown (Comins & Trenbath, to be published) that the "equilibrium" strategy generally gives minimum selection. If the optimal yielding strategy is "min-max" it may be necessary to derive a compromise strategy by assigning an equivalent short term cost to resistance-breaking.

ACKNOWLEDGEMENTS

Hugh Comins is supported by a Queen Elizabeth II fellowship.

REFERENCES

Bellman, R.E. 1957. Dynamic Programming, Princeton University Press, Princeton, N.J.

Comins, H.N. and Trenbath, B.R. 1980. Optimal Strategies for the Subinjurious Maintenance of Nematode Populations, to be published

de Wit, C.T. 1960. Versl. Landbouwk. Onderz. Ned. 66, 1-82

Hassell, M.P. 1975. Density-dependence in Single-species Populations, J. Anim. Ecol. 44, 283-295

Nusbaum, C.J. and Barker, K.R. 1971. Population Dynamics, in Plant Parasitic Nematodes, ed. Zuckerman, B.M., Mai, W.F. and Rohde, R.A., Acad. Press, N.Y.

Sibma, L., Kort, J. and de Wit, C.T. 1964. Jaarb. Inst. Biol. Scheik. Onderz. Landbouwk., Wageningen, 119-124

Taylor, D.P. 1976. Plant Nematology Problems in Tropical Africa, Helminth. Abstracts 45, 269-284

Trenbath, B.R. 1977. Interactions among Diverse Hosts and Diverse Parasites, Annals of the New York Academy of Sciences 287, 124-150

A PLANT-WATER MODEL WITH IMPLICATIONS FOR
THE MANAGEMENT OF WATER CATCHMENTS

R. McMurtrie
Division of Forest Research, CSIRO
P.O. Box 4008
Canberra   ACT 2600
Australia

An understanding of plant-water relationships is essential to the sound
management of water catchments. This paper describes a general model of
how the availability of soil-water affects the photosynthetic growth of
plants. Factors affecting the water status of a plant include precipitation, soil drainage, evaporation, run-off, water uptake by vegetation
and transpiration. These factors are combined in a differential equation model to derive a relatively simple, graphical criterion for the
survival of a plant. The model has implications for the management of
water catchments and could provide a helpful approach to the study of
the successional process and of inter-plant competition for soil water.

INTRODUCTION

Mountain ash (Eucalyptus regnans) forests make up approximately 50 per cent of
the total catchment area (of 120 000 ha) supplying water to the city of Melbourne
and yield 70 to 80 per cent of the city's water supply. Mountain ash seedlings are
intolerant to shading and regeneration occurs in nature after wildfires of sufficient intensity to remove both the overstorey and understorey. Large amounts of
seed are released by the scorched canopy and regrowth following a wildfire is both
rapid and dense. Trees typically reach heights of 20 m in a 5 or 6 year period.

In 1939 a crownfire swept through most of Melbourne's catchment area. The
open mature mountain ash forests were replaced by dense regrowth. The regrowth
forests have a much higher demand for water than forests of mature trees. A marked
reduction in catchment water yield is apparent in streamflow data less than 5 years
after the fire with an estimated water loss of 25% over the 20 year period leading
up to 1964 in catchments where mountain ash is the dominant overstorey species.
The evidence is that the water yield is now gradually returning to prefire levels.
In catchments where mountain ash is not the dominant overstorey species the decline
in water yield does not appear to have occurred. (See Langford (1974) for a detailed exposition of the impact of the 1939 bushfire on catchment yield of water.)

It is within this context that it is particularly relevant to construct models
relating plant growth and water use. Section 2 describes a general model of how
the photosynthetic growth of a plant depends upon the availability of soil water in
relation to the plant's demand for water. Important components of the water balance

include precipitation, soil drainage, evaporation, water uptake by vegetation and transpiration. These factors are combined in a model of photosynthetic growth both to derive a relatively simple, graphical criterion for the survival of a seedling under conditions of water stress and to provide a model to predict plant growth and water use.

The model considers the relationship between three dynamic variables, the weight of the plant, the leaf water potential and the soil water potential. Plant growth occurs if photosynthetic input is surplus to the plant's maintenance requirement. In the model the rate of photosynthetic input varies as stomata open and close in response to changes in leaf water potential. Changes in leaf water potential are determined by a balance between the plant's water uptake, which depends upon the difference between leaf water potential and soil water potential, and transpiration loss which varies with leaf water potential due to opening and closing of stomata. Soil water potential varies depending upon throughfall input and losses due to uptake by vegetation, drainage and evaporation.

The model leads to a criterion for a seedling to grow or to die. According to the criterion, the following changes in model parameters are conducive to seedling survival: an increased throughfall rate (throughfall is rainfall minus interception by vegetation); an increase in the seedling's rate of uptake of moisture from the soil; increased radiation level; and increase in the critical value of leaf water potential below which transpiration declines; a decrease in the plant's maintenance requirement; reduced drainage loss from the soil; reduced transpiration loss; reduced uptake by competing vegetation; and a decrease in the critical value of leaf water potential below which photosynthesis declines.

The criterion offers an insight into how the growth of one plant species can be inhibited by competition for water from a second species. The model could provide an approach to the stu y of the successional process. The fact that an early successional species disappears might be explainable in terms of changes in model parameters (e.g., corresponding to increased water uptake by competing vegetation or to specific adaptations of higher successional species to increase their own fitness at the expense of competitors) meaning that the species' seedlings can no longer survive. By making statements about the survival of seedlings, the model might also shed light on the nature of boundaries and gradients in the spatial distribution of vegetation. The model could also be used to predict optimal strategies for stomatal closure. While only water cycling is considered in this model, there is no reason why the cycling of other nutrients could not be accommodated in a modelling framework similar to the present model of water cycling.

By linking the dynamics of water cycling and the dynamics of vegetation growth, models of this kind might also be a helpful guide to the management of water catchments. Section 3 constructs a qualitative model of how the pattern of water use might change as the forest regrows after clearfelling or after devastation by fire. A fourth dynamic equation is introduced to the single plant model to represent changes in the number of stems per unit area. In the model stem density increases linearly with time following regeneration until self thinning comes into effect. The model is simulated to monitor changes in soil drainage with regrowth. In the early stages of regrowth, soil drainage is high because the regenerating forest consumes relatively little water. However, as both stem number and the water uptake of an individual plant increase, water consumption by vegetation increases leading to a decline in catchment discharge. The pattern in soil drainage as the forest continues to mature depends upon whether the product of the number of surviving stems and the water uptake by an individual tree decreases or continues to increase with forest age. For two models simulated in Section 3, soil drainage follows a pattern similar to that observed for streamflow observations in the mountain ash forests of the Melbourne catchment area following the 1939 wildfire. Here, a short-term increase in drainage is followed by a steady decline to a level well below that of the mature forest, before drainage begins a gradual return to its initial level.

GROWTH-WATER MODEL FOR A SINGLE PLANT

A common approach in modelling the physiology of plants is to treat plant growth as a process which occurs when photosynthetic input is surplus to a plant's maintenance requirements (Thornley 1970, 1971, Beevers 1970, deWit, Brouwer and Penning de Vries 1970, Ross 1970, Hesketh, Baker and Duncan 1971, Jones and Smerage 1978). Consider a plant of total weight W. Several studies propose models of the form

$$\frac{dW}{dt} = Y \ (P - M) \tag{1}$$

where P is the rate of gross photosynthetic input of substrate, where M represents the loss of substrate through maintenance respiration and where Y is an efficiency factor. Both P and M would in general be functions of W and of environmental variables,

$$P = p(W(t))$$

and $$M = m(W(t)) \ . \tag{2}$$

At time t an increase in plant weight occurs if P exceeds M. The form of the dependence of P on W is discussed by Ledig (1969) and Thornley (1976, 1977). It is common practice to assume that m(W) is proportional to W, (e.g., Hesketh, Baker and

Duncan 1971). Let ρ and μ represent the values assumed by $(\frac{\partial p}{\partial W})$ amd $(\frac{\partial m}{\partial W})$ respectively, in the limit as W approaches zero.

The implications of this model for the survival of a seedling can be understood by considering the stability of the equilibrium W=0. (Note that p(0)=0 and m(0)=0.) Depending on the parameter values, the equilibrium is either stable, corresponding to a dying plant or unstable, corresponding to a growing plant. The plant grows if

$$r = \mu/\rho \quad < 1 \quad . \tag{3}$$

The survival prospects of a seedling are determined by a balance between the photosynthetic input and the maintenance requirement at low values of W. Effectively, death occurs if the radiation level is not sufficient to meet the plant's maintenance requirement.

This section extends the model (1) to consider how survival is affected by the availability of soil water. In the model I assume that plant water status affects growth through its influence on the rates of water uptake, transpiration and photosynthesis. The reduction in photosynthesis and transpiration under conditions of water deficit is usually attributed to reduced rates of CO_2 exchange resulting from stomatal closure and decreases in leaf surface area and to direct effects upon biochemical processes (Slatyer 1967, Crafts 1968 and Kramer 1969). Figure 1 illustrates how the rates of photosynthesis P and transpiration T might depend upon the leaf water potential ψ_L (functions f and g respectively):

$$P = p(W)f(\psi_L) \text{ and } T = bg(\psi_L) \tag{4}$$

where b is a constant representing the level of transpiration when water is abundant. Environmental factors such as atmospheric temperature and humidity are implicit in the value of b. When water is abundant the expression for photosynthesis reduces to (2). The functions f and g represent the proportional reduction in photosynthesis and transpiration under conditions of water deficit.

The value of ψ_L for a plant will itself vary depending upon uptake from the soil and loss through transpiration. The rate of water movement through the plant is generally assumed to be proportional to the local gradient in water potential (e.g., Taylor 1968). This assumption implies that the rate of movement of water from soil to leaf is approximately proportional to the difference between the water potentials of the soil and the leaf (Woo, Boersma and Stone 1966). Jones and Smerage (1978) and Woo et al. (1966) assume that fluctuations in ψ_L are proportional to fluctuations in leaf water content (LWC). (The pros and cons of assuming a linearly relationship between ψ_L and LWC are spelled out by Barrs (1968)).

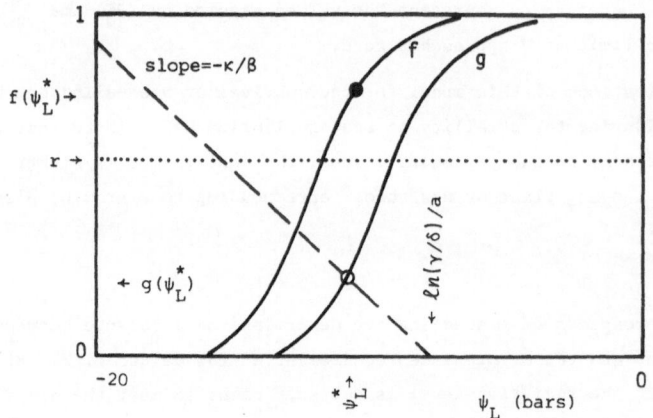

FIGURE 1. Graphical solution of the survival criterion (9). The equilibrium solution of (8) is circled. The rates of photosynthesis and transpiration are represented by functions f and g, respectively. The seedling survives since the value of $f(\psi_L^*)$, represented by the dot, lies above the dotted line.

The leaf water potential satisfies an equation of the form

$$\frac{d}{dt} (\psi_L) = \kappa(\psi_L - \psi_S) - \beta g(\psi_L) \tag{5}$$

where ψ_S is the soil water potential measured in the rooting zone and where κ and β are constants corresponding to the rates of water uptake from the soil and transpiration loss to the atmosphere, respectively.

Several factors affect soil water content (SWC) including throughfall, water uptake by vegetation, evaporation, drainage and run-off. Assume initially that the bulk of soil water loss occurs through drainage. The evidence from Fleming and Smiles (1975) is that drainage D scales as the exponential of soil water potential. Assume

$$D = \delta \exp (a \psi_S) . \tag{6}$$

Moisture is effectively drained when it is lost from the rooting zone. In the expression (6) a and δ are constants and drainage approaches zero at low SWC. The value of δ corresponds to the drainage rate when $\psi_S = 0$. Assume that fluctuations in ψ_S are proportional to fluctuations in SWC (Woo et al. (1966)) to derive a model for ψ_S

$$\frac{d}{dt} (\psi_S) = \gamma - \delta exp(a\psi_S) \tag{7}$$

where γ corresponds to the rate of throughfall and where the exponential term corresponds to drainage losses D.

The model (1,5,7) possesses an equilibrium W^*, ψ_L^*, ψ_S^*

where

$$W^* = 0$$

$$\psi_S^* = \frac{1}{a} \ln (\gamma/\delta)$$

and where ψ_L^* is solution of the equation

$$\frac{\kappa}{a\beta} \ln(\gamma/\delta) - \frac{\kappa}{\beta} \psi_L^* = g(\psi_L^*) \quad . \tag{8}$$

If the equilibrium is stable the seedling does not grow since the solution is attracted to the zero state. Neighbourhood stability analysis is performed by linearising the model (1,5,7) about the equilibrium (8). The seedling survives if the equilibrium is unstable which follows if

$$f(\psi_L^*) > r \quad . \tag{9}$$

Note that for the plant to grow under conditions of optimal moisture requires that $r < 1$ (from (3)) and that the value of $f(\psi_L^*)$ lies between 0 and 1.

Equations (8) and (9) can be combined to provide a graphical representation of the survival criterion. See Fig. 1. Equation (8) can be solved graphically as the intersection of $g(\psi_L^*)$ with the dashed straight line illustrated. The plant grows only if the relative photosynthesis $f(\psi_L^*)$ exceeds a constant value r which is itself less than unity. The value r is represented in Fig. 1 by the dotted horizontal line. In Fig. 1 where (9) is obeyed the plant survives. Death occurs if the value of $f(\psi_L^*)$ lies below this line. (Note that the graphical analysis applies even if it is not justified to assume that fluctuations in water potential are proportional to fluctuations in water contents.)

Although it is unrealistic to assume that model parameters are time invariant, the model can be used as a guide to understanding the determinants of survival. Variations in parameter values could mean that the solution of (1) is sometimes increasing and sometimes decreasing. To assess the effect of temporal variability in parameters, the model could be used to estimate the proportion of days on which P exceeds M. In practice the assumption of time invariant parameters means only that parameter fluctuations are small and occur over a time scale short compared to other time scales in the model and that it is possible to average over the fluctuations. Fluctuations or cyclic changes occurring on a longer time scale could be accommodated in a simulation model. Since the water potentials vary on a timescale much faster than the plant weight, it might be helpful to decouple the dynamics of water transport from those of plant growth using the techniques described by Ludwig, Jones and Holling (1978). This approach could provide a guide to how the growth rate varies with fluctuations in some environmental parameters.

The graphical representation of the survival criterion offers a straight-forward interpretation of how model parameters affect the survival prospects of seedlings. An increase in the radiation level or a decrease in the value of μ means a lowering of the dotted line in Figures 1 and 2, and hence will promote survival.

In Fig. 2(a) the effect of increasing the ratio (γ/δ) is to raise the value of $f(\psi_L^*)$. Thus increased rainfall or reduced drainage loss from soil is conducive to survival. Figure 2(b) illustrates the dependence on the ratio (κ/β). Clearly, an increase in water uptake by the seedling or a reduction in the maximum transpiration rate tends to enhance the seeling's survival prospects.

The shapes of the functions $f(\psi_L)$ and $g(\psi_L)$ also hold implications for the survival of seedlings. Figure 2(c) compares seedlings with and without the ability to photosynthesise at low ψ_L. The former are more likely to survive. Barrs (1968) argues that transpiration and photosynthesis vary in unison with variations of leaf water potential. If it is valid to assume that $f(\psi_L^*) = g(\psi_L^*)$ then the graphical approach is greatly simplified. Slatyer (1967) on the other hand, suggests that stomatal closure will reduce transpiration more than photosynthesis.

Of course, common sense, qualitative predictions such as those above would be expected of any sound model. However, it is interesting that statements on relationships between water cycling and survival arise from a simple theoretical model. Because of the graphical nature of the survival criterion the model could readily be applied to controlled environment experiments. The survival criterion derived in this paper is not specific to models of the form (1). For instance a criterion with a similar graphical representation emerges for the storage model of Thornley (1977).

In addition to making statements on seedling survival, models of the form (1)

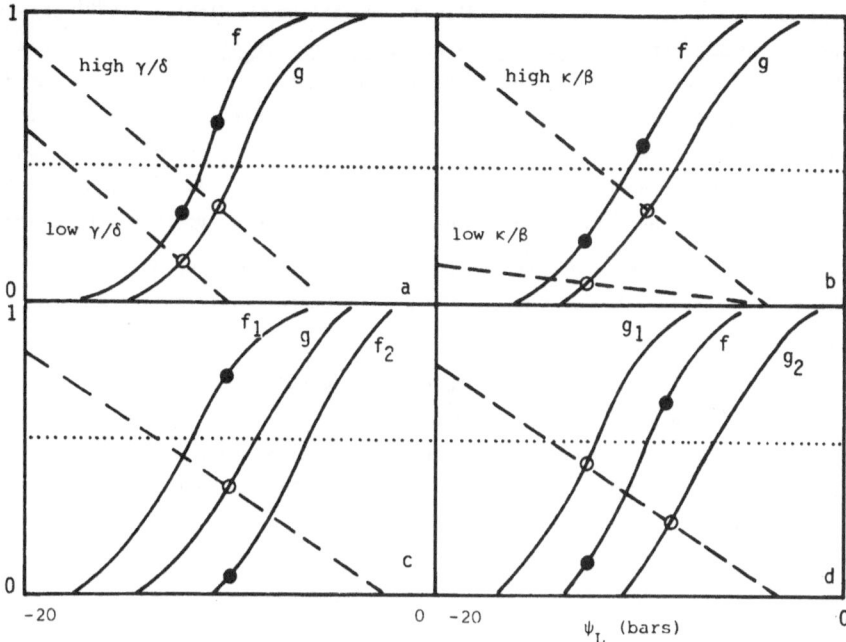

FIGURE 2. The relationship between survival and the value of parameters in the model. A seedling is more likely to survive if

      a)   it has a relatively high value of the ratio (γ/δ) ,

      b)   it has a relatively high value of the ratio (κ/β) ,

      c)   it is able to photosynthesise at relatively low $\psi_L(f_1)$,

      d)   it ceases transpiration at relatively high $\psi_L(g_2)$.

could be used to predict plant growth. For instance the dominant eigenvalue of the stability matrix for the equilibrium (8) of the differential equation model (1,5,7) provides a simple expression for the growth rate of a seedling.

The model offers some understanding of how interspecific competition for soil water can affect survival. Whereas in (7) the value of ψ_S^* is determined by the balance between throughfall and drainage, when competition between plants for soil moisture is intense, the balance will be struck between throughfall and the sum of drainage and transpiration by the seedling's competitors. Withers (1979) observed that when seedlings of Eucalyptus ovata and Casuarina littoralis are grown and droughted in competition, the higher rate of water uptake by E. ovata subjected C. littoralis to greater stress and earlier death than it experienced in monoculture. A graphical argument can clarify this result. The effect of introducing a competitor with rapid water uptake is to enhance the loss of soil water in Equation (7) and hence to reduce the equilibrium value of ψ_S. This tends to lower the dashed line of Figs. 1 and 2, making survival less likely. Other experiments have reported similar competition effects (Bunce, Miller and Chabot 1977 and Hartmann and Allard 1964).

Dougherty (1972) cites an example where root development of red clover is restricted to within a few cm of the surface when it is grown in association with wheat. The red clover is under considerable water stress even when soil water is abundant. In terms of the present model competition has the effect of reducing the uptake rate κ of red clover.

The criterion (9) also offers an insight into how competition for soil water might limit the density of vegetation. As vegetation grows more dense, new seedlings suffer greater stress perhaps due to the decline in SWC resulting from the increased water uptake of vegetation and from increased interception of rainfall, or perhaps due to a decline in ρ caused by increased shading by larger plants. Associated with the decline is a gradual lowering of the dashed and dotted lines of Figures 1 and 2 until the stage is reached where new seedlings are unable to survive.

GROWTH-WATER MODEL FOR A FORESTED CATCHMENT

Difficulties arise in attempting to extend the above seedling model to describe the growth dynamics of mature plants because of variations in parameter values with age. However it is possible to draw some tentative conclusions based on the assumption of time invariant parameters. The purpose of this section is to show that the pattern of stream discharge observed in the mountain ash forests of the Melbourne catchment area following the 1939 bushfire can be captured by the simple model constructed above. The model described below is built on tenuous assumptions and is intended only to make armwaving, preliminary statements about possible explanations of the behaviour of the mountain ash water catchments.

Let W represent the average weight of a tree and let N represent the number of living trees. Invoking model (1) above, assume that W obeys the equation

$$\frac{dW}{dt} = Y(p(W)f(\psi_L) - m(W)) \tag{10}$$

where the rate of photosynthetic input is

$$p(W) = \frac{\rho W}{1+bW} \, , \tag{11}$$

where

$$f(\psi_L) = \frac{1}{1+\psi_L/\psi_{Lp}} \tag{12}$$

and where the maintenance requirement is proportional to W (Hesketh, Baker and Duncan (1971))

$$m(W) = \mu W \quad . \tag{13}$$

Parameters Y, P, b, μ, ψ_{Lp} are constants. Here the photosynthetic rate is a linear function of W for a regrowth forest and approaches an asymptote as the forest matures.

Consider a forest destroyed by fire at time t=0. Assume that seedling regeneration begins immediately, with stem number N increasing linearly with time. Stem number continues to rise until the carrying capacity of the environment is reached. The stem number then declines naturally through self-thinning. According to the 3/2 power law of self-thinning (Yoda, Kira, Ogawa and Hozumi 1963 and Miyanishi, Hoy and Cavers 1979), the average weight of surviving trees scales inversely as the 3/2 power of the number of surviving stems. (A plausability argument leading to this law can be presented in terms of the dimensional relationship between the weight per plant, scaling as the cube of a linear dimension, and the ground area covered per plant, scaling as the square of a linear dimension.)

Assume that the ground area covered by a tree of weight W is $vW^{2/3}$ where v is a constant and that the stem number increases if the total coverage $vNW^{2/3}$ is less than the total available space s. Otherwise the stem number declines. A differential equation which encapsulates this behaviour is

$$\frac{dN}{dt} = s - v \, NW^{2/3} \quad . \tag{14}$$

The model (14) embodies an initial linear rate of increase and its solutions approximately follow the 3/2 law of self-thinning.

The forest's rate of water uptake is a function of stem number and of average tree weight. Assume that the rate of reduction of soil water potential due to uptake by the forest is

$$U_S = N \, h_S(W) \, (\psi_S - \psi_L) \quad . \tag{15}$$

Note that it is probably reasonable to assume that all mountain ash trees are of equal W since a mountain ash forest, regrowing after fire, will be approximately even-aged. The dynamics of soil water potential is

$$\frac{d\psi_S}{dt} = \gamma - D - U_S \tag{16}$$

where the throughfall rate is represented by γ and where drainage D is given by (6).

The leaf water potential varies depending upon uptake from the soil and tran-spiration loss. Assume that the rate of increase of leaf water potential due to uptake from the soil is

$$U_L = h_L(W) \ (\psi_S - \psi_L) \tag{17}$$

The dynamics of leaf water potential is

$$\frac{d\psi_L}{dt} = U_L - \beta \ g \ (\psi_L) \quad . \tag{18}$$

where the transpiration term is of the form

$$g(\psi_L) = \frac{1}{1+\psi_L/\psi_{Lt}} \tag{19}$$

The functions h_L and h_S will be related, with h_S representing a water flux and h_L representing a flux per unit leaf volume.

One factor which will be reflected in the W dependence of h_S and h_L is how the leaf area of a tree varies with W. Assume that the rate of water uptake by an individual tree is proportional to its leaf area which is assumed to scale as $W^{2/3}$. Consider the simulations of Fig. 3 for which

$$U_L = \kappa \ (\psi_S - \psi_L) \tag{20}$$

$$U_S = cNW^{2/3} \ (\psi_S - \psi_L) \quad . \tag{21}$$

Notice in Fig. 3 the decrease in drainage from age 2 to age 9 years followed by a gradual increase. This is similar to the behaviour of the mountain ash forests of the Melbourne catchment area after the 1939 wildfire.

In the model the scaling of U_L and U_S with increasing W is critical in deter-mining the trend in water use. Note that according to the 3/2 law of self-thinning,

$NW^{2/3}$ approaches a constant as the forest matures.

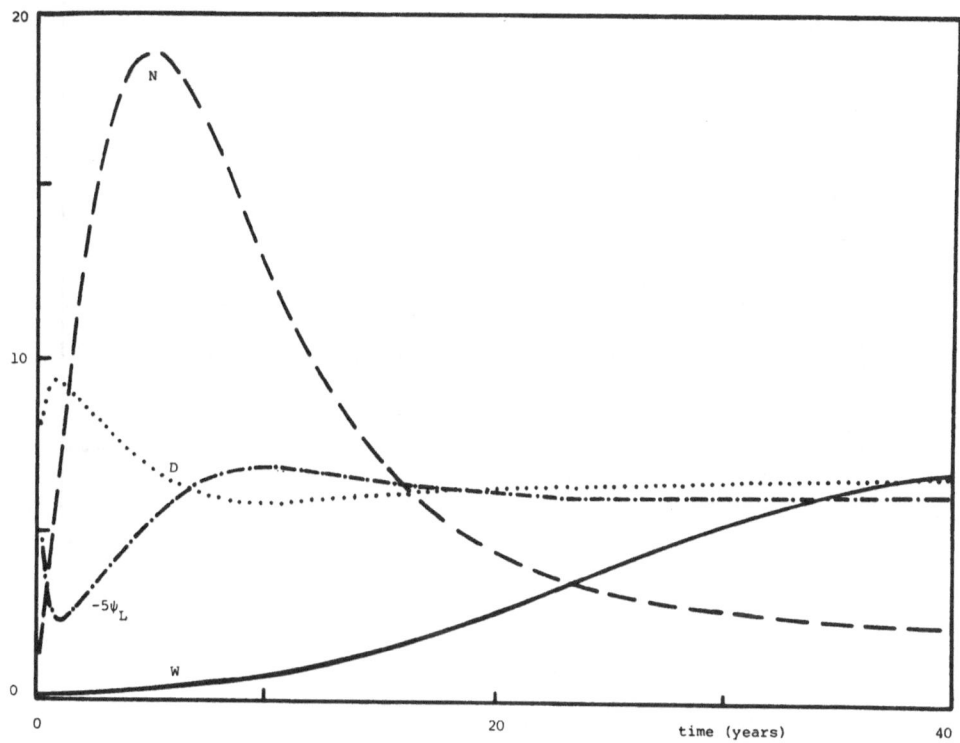

FIGURE 3. Simulation of the catchment model (10,11,12,13,14,16,18,19,20,21) . Parameter values are Y=0.5, ρ=1, b=0.1, μ=0.5, s=8, v=1, κ=6, β=2, γ=10, δ=10, a=0.5, c=1.5, ψ_{Lp}= -10, ψ_{Lt}= -10 . Time is measured in units of years.

The vertical axis is scaled in arbitrary units. Notice that the mature forest differs from the regrowth forest in its lower water consumption and higher drainage. The number of surviving trees is represented by N, the average weight of surviving trees by W, the soil drainage by D, and the leaf water potential by ψ_L .

The simulations of Fig. 4 assume that

$$U_L = \kappa \ (\psi_S - \psi_L) \tag{22}$$

$$U_S = cNW^{1/2} \ (\psi_S - \psi_L) \ . \tag{23}$$

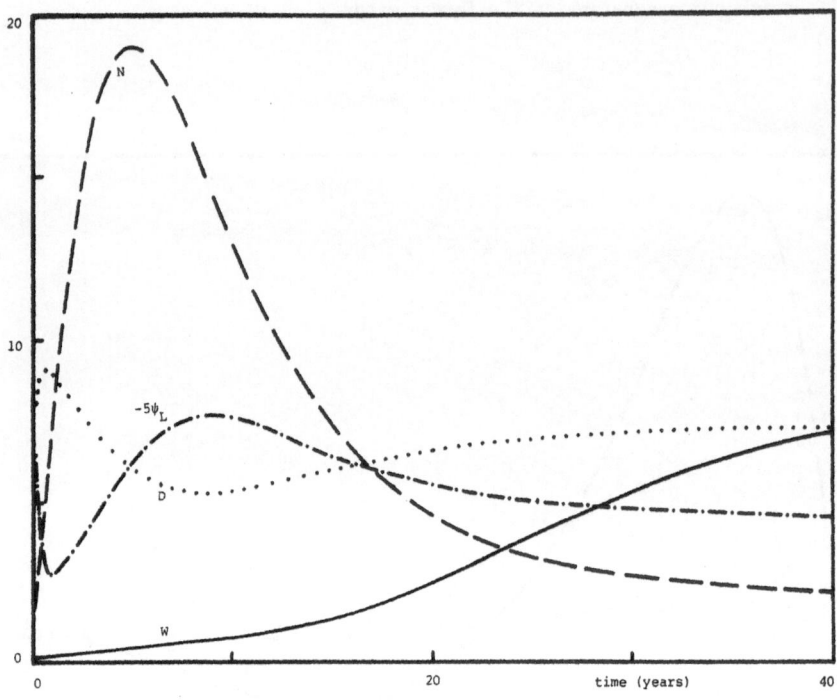

FIGURE 4. Simulation of the model (10,11,12,13,14,16,18,19,22,23) with parameter values of Figure 3 .

Provided $NW^{2/3}$ is approximately constant, the total water uptake (23) decreases as W increases, with a consequent increase in discharge of water. The simulations of Figure 5 assume that

$$U_L = \kappa \ (\psi_S - \psi_L) \tag{24}$$

$$U_S = cNW^{5/6} \ (\psi_S - \psi_L) \ . \tag{25}$$

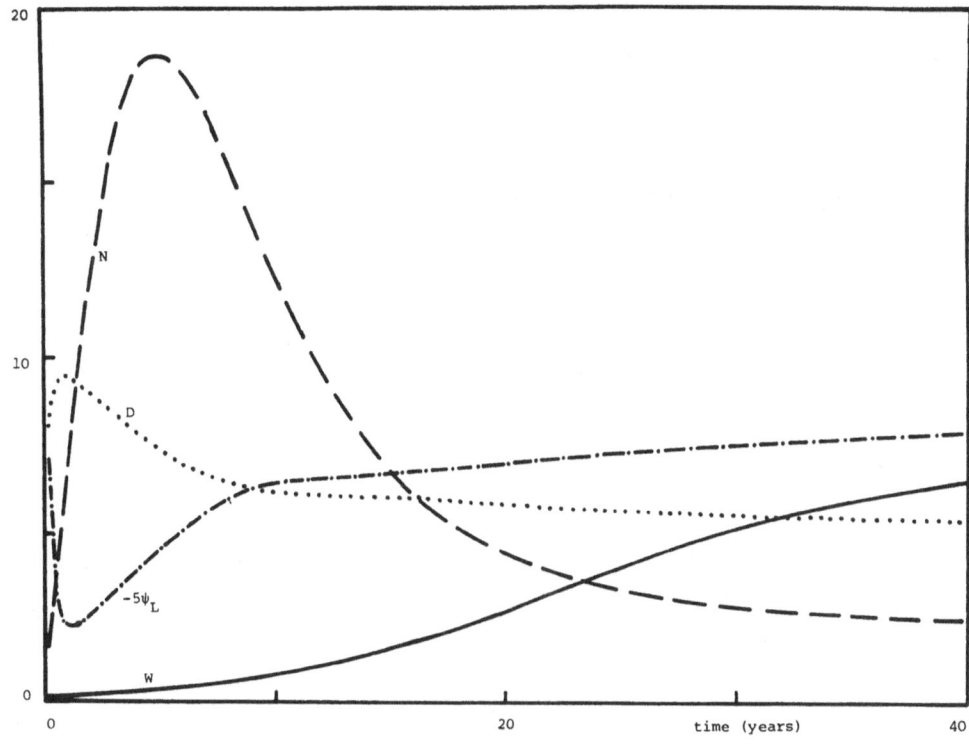

FIGURE 5.   Simulation of the model (10,11,12,13,14,16,18,19,24,25) with parameter
values of Figure 3 .

Under the 3/2 power law, the total rate of water uptake by the forest (25) increases as the forest matures, with a corresponding decrease in water discharge.
The dynamic behaviour of Figure 5 is in contrast to that of the mountain ash
catchments.

Langford (1974) considers that changes in water consumption by the mountain
ash catchments probably arise from changes in the transpiration rate caused by a
lowering of leaf water potential as tree height increases.   An effect of height
growth is to introduce an additional potential ψ_g due to gravitation in the uptake
rates (15) and (17),   The simulations of Figure 6 consider functions U_L and U_S of

the form

$$U_L = \kappa \ (\psi_S - \psi_L - \psi_g) \qquad\qquad (26)$$

$$U_S = cNW^{2/3} \ (\psi_S - \psi_L - \psi_g) \qquad\qquad (27)$$

where ψ_g which is proportional to tree height is assumed to scale as the cube root of tree weight

$$\psi_g = dW^{1/3} \ . \qquad\qquad (28)$$

Figure 6 simulates the model (10,11,13,14,16,18,26,27) with $f(\psi_L)$ and $g(\psi_L)$ step functions of ψ_L:

$$f(\psi_L) = 0 \quad \text{if} \quad \psi_L < \psi_{Lp}$$

$$f(\psi_L) = 1 \quad \text{if} \quad \psi_L > \psi_{Lp} \qquad\qquad (29)$$

and

$$g(\psi_L) = 0 \quad \text{if} \quad \psi_L < \psi_{Lt}$$

$$g(\psi_L) = 1 \quad \text{if} \quad \psi_L > \psi_{Lt} \ . \qquad\qquad (30)$$

Figure 6(a) simulates the model with $\psi_g = 0$ and Figure 6(b) with ψ_g given by (28).

The dynamic behaviour of the model is consistent with Langford's (1974) comment on why the observed pattern of water discharge could occur. The exploratory simulations of Figures 3,4,5 and 6 illustrate that the above model could serve as a tool for testing hypotheses concerning the causes of differences in water consumption by regrowth and mature mountain ash forests.

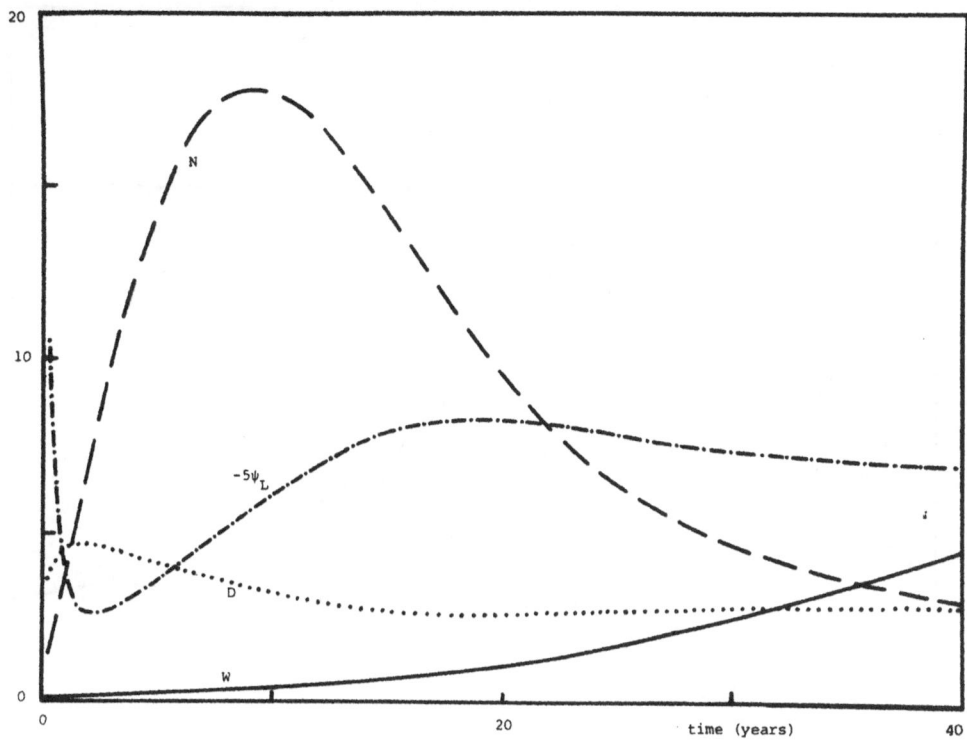

FIGURE 6. Simulation of the model (10,11,13,14,16,18,26,27,29,30)

(a) ignoring height effect $\psi_g = 0$.

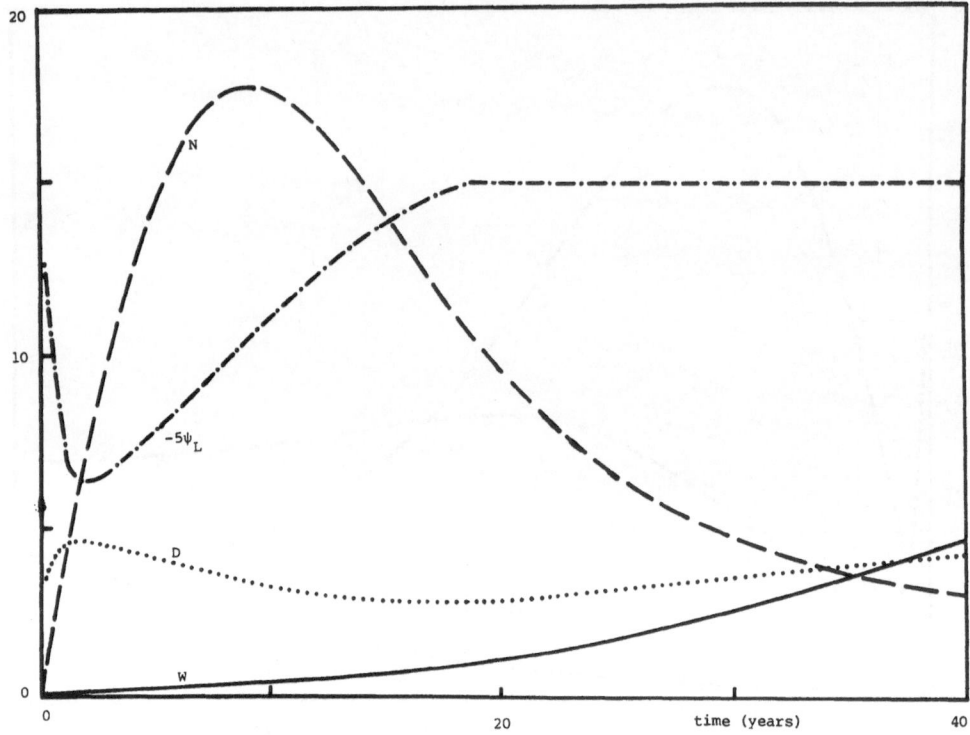

FIGURE 6.   Simulation of the model (10,11,13,14,16,18,26,27,29,30)

(b) with ψ_g given by (28). As the forest matures height growth tends to reduce leaf water potential. Transpiration losses continue until ψ_L reaches the critical value ψ_{Lt}. A discontinuity in ψ_L and ψ_S occurs at this point because the transpiration rate is no longer maintained at its maximum. In (18) water uptake decreases to match the reduced transpiration rate. There is a consequent increase in drainage from the soil. Parameter values are Y=0.5, ρ=0.5, b=0.1, μ=0.25, s=4, v=0.5, κ=3, β=1, γ=5, δ=5, a=0.5, c=0.75, ψ_{Lp} = -10, ψ_{Lt} = -3. In Figure 6(b) d = 1.5 .

REFERENCES

Barrs, H.D. 1968. Determination of water deficits in plant tissues. In Water Deficits and Plant Growth, Vol. 1, T.T. Kozlowski (Ed.), Academic, New York, Ch. 8

Beevers, H. 1970. Respiration of plants and its regulation. In Prediction and Measurement of Photosynthetic Productivity, I. Setlick (Ed.), Pudoc, Wageningen, 209-214

Bunce, J.A., Miller, L.N., and Chabot, B.F. 1977. Competitive exploitation of soil water by five eastern North American tree species. Bot. Gaz. 138, 168-173

Crafts, A.S. 1968. Water deficits and physiological processes. In Water Deficits and Plant Growth, Vol. 2, T.T. Kozlowski (Ed.), Academic, New York, Ch. 3

deWit, C.T., Brouwer, R., and Penning de Vries, F.T. 1970. The simulation of photosynthetic systems. In I. Setlick (Ed.), ibid, 47-70

Dougherty, C.T. 1972. Water stress in Turoa red clover under Aotea wheat. N.Z. J. Agric. Res. 15, 706-711

Fleming, P.M. and Smiles, D.E. 1975. Infiltration of water into soil. In Prediction in Catchment Hydrology, T.G. Chapman and F.X. Dunin (Eds) (Aust. Acad. Sci.) 83-110

Hartmann, R.W. and Allard, R.W. 1964. Effect of nutrient and moisture levels on competitive ability in barley (Hordeum vulgare L.) Crop. Sci. 4, 424-426

Hesketh, J.D., Baker, D.N., and Duncan, W.G. 1971. Simulation of growth and yield in cotton : respiration and the carbon balance. Crop. Sci. 11, 394-398

Jones, J.W. and Smerage, G.H. 1978. Representation of plant-crop physiology. ASAE Tech. Paper No. 78-4024

Kramer, P.J. 1969. Plant and Soil Water Relationships: A Modern Synthesis, McGraw-Hill, New York

Langford, K.J. 1974. Change in Yield of Water Following a Bushfire in a Forest of Eucalyptus regnans, Water Supply Catchment Hydrology Research, Melbourne and Metropolitan Board of Works. Report No. MMWB-W-0003

Ledig, F.T. 1969. A growth model for tree seedlings based on the rate of photosynthesis and the distribution of photosynthate. Photosynthetica 3, 263-275

Ludwig, D., Jones, D.D., and Holling, C.S. 1978. Qualitative analysis of insect outbreak systems : the spruce budworm and forest. J. Anim. Ecol. 47, 315-332

McCree, K.J. 1970. An equation for the rate of respiration of white clover plants grown under controlled conditions. In I. Setlick (Ed.), ibid., 221-229

Miyanishi, K., Hoy, A.R., and Cavers, P.V. 1979. A generalized law of self-thinning in plant populations. J. Theor. Biol. 78, 439-442

Ross, J. 1970. Mathematical models of photosynthesis in a plant stand. In I. Setlick (Ed.), ibid., 29-45

Slatyer, R.O. 1967. Plant-Water Relationships, Academic, London

Taylor, S.A. 1968. Terminology in plant and soil water relations. In Water Deficits and Plant Growth, Vol. 1, T.T. Kozlowski (Ed.), Academic, New York, Ch. 3

Thornley, J.H.M. 1970. Respiration growth and maintenance. Nature, London 225, 304-305

Thornley, J.H.M. 1971. Energy, respiration and growth in plants. Ann. Bot. 35, 721-728

Thornley, J.H.M. 1976. Mathematical Models in Plant Physiology, Academic, London

Thornley, J.H.M. 1977. Growth maintenance and respiration : a reinterpretation. Ann. Bot. 41, 1191-1203

Withers, J.R. 1979. Studies on the status of unburnt Eucalyptus woodland at Ocean Grove, Victoria. The interactive effects of droughting and shading on seedlings under competition. Aust. J. Bot. 27, 285-300

Woo, K.B., Boersma, L., and Stone, L.N. 1966. Dynamic simulation model of the transpiration process. Water Resources Research, 2, 85-97

Yoda, K., Kira, T., Ogawa H., and Hozumi, K. 1963. Self-thinning in overcrowded pure stands under cultivated and natural conditions (intraspecific competition among higher plants XI). J. Biol. Osaka City Univ. 14, 107-129

MATHEMATICAL MODELLING OF THE TRANSPORT AND LOSS OF
LEACHABLE PLANT NUTRIENTS
IN FIELD SOILS

C. W. Rose
School of Australian Environmental Studies
Griffith University, Nathan
Queensland, 4111, Australia

The paper describes a limited range of mathematical models, and field
applications of these models, for one-dimensional transport of mobile
(lightly sorbed) nutrients in soils. A model appropriate for use in
soils which drain to a 'field capacity' value is given which allows
the movement in peak solute concentration ot be predicted from any
series of irrigation/rainfall and evapotranspiration events. Agreement with the field data of Saffigna (1977) is illustrated. A steady-
state solution which uses this peak solute position to account for
mass flow or convection was then used to provide a model which also
describes the effect of dispersion on solute concentration. This model
was then applied to the ^{15}N-labelled fertilizer study of Chichester
and Smith (1978) in field monolith lysimeters. The approach provided
a relatively simple and accurate method of predicting nitrate concentrations as a function of depth and time allowing for the effects of
mass flow, dispersion, and (separately measured or estimated) nitrogen
uptake and transformation. The method is approximate but analytic in
character.

INTRODUCTION

Nutrients essential to the functioning of natural or man-managed ecosystems
vary enormously in the degree to which they form compounds which are readily transported through the soil profile. These nutrients may be roughly classified into
those which do, and those which do not have very soluble and therefore leachable
forms. Nitrogen and potassium are examples of mobile, leachable nutrients;
phosphorus is an example of a strongly sorbed non-leachable nutrient.

All resource development appears to suffer to some extent from unintended and
often undesired consequences, frequently referred to as environmental problems.
For example, on average about half the nitrogenous fertilizer applied is never
taken up by the crop for which it was intended. Much of this loss that can potentially be reduced by management practices is due to 'leaching' of this mobile ion
beneath the root zone, leading to economic loss and subsequent deterioration in
water quality.

Controlling the loss of strongly-sorbed nutrients, on the other hand, devolves
to conserving from erosive loss the soil to which such nutrients are bound, whether
such erosion is induced by water or wind.

There is very considerable potential for the use of mathematical models to aid the development of management practices which minimise the undesired consequences accompanying resource development.

This paper illustrates the mathematical (and to some extent the physical) basis of relatively simple, management-oriented models of mobile solute movement in soils. The capacity of models to interpret the movement of mobile plant nutrients in the field is illustrated.

The theory presented is applicable not only to mobile plant nutrients, but also to lightly-sorbed salts such as sodium chloride which, at sufficient concentration, can be deleterious to plant growth and aggravate soil erosion.

In what follows let c denote the concentration of any non-sorbed and non-precipitating solute, Θ the volumetric water content, and q the volume flux density of solution in the z-direction. Assume further, for the present, that the solute is transported purely by mass flow or convection and is not taken up by plants. Under these assumptions, conservation of mass of solute requires that

$$\frac{\partial(\Theta c)}{\partial t} = - \frac{\partial(qc)}{\partial z} \tag{1}$$

where t denotes time.

SALT DYNAMICS IN IRRIGATED SLOWLY-PERMEABLE SOILS

Modest plant uptake of sodium chloride can allow direct use to be made of Eq. (1) to investigate salt dynamics under irrigation, as shown by Rose et al. (1979). Suppose field sampling to depth D gives spatial mean NaCl concentrations $\bar{c}$ (t_1) at time t_1, and $\bar{c}$ (t_2) at some later time t_2. Leaching of salt takes place chiefly when water content is near some higher value, denoted $\bar{\Theta}_f$.

Integrating Eq. (1) from the soil surface (z = 0) to z = D, with the implicit assumptions of the above paragraph gives:

$$D\bar{\Theta}_f \frac{\partial \bar{c}}{\partial t} = q_o c_o - q_D c_D \tag{2}$$

where $q_o c_o$ represents the flux density of salt at the soil surface, arriving through irrigation water and rainfall, for example.

Experimental data can show (Rose et al. 1979) that over a time interval $(t_1 \rightarrow t_2)$ in impermeable soils:

$$c_D = \lambda c_o \qquad (3)$$

where λ is approximately constant.

Equation (2) can then be solved to give:

$$\bar{c}(t_2) - \bar{c}(t_1) = [(q_o c_o/q_D \lambda - \bar{c}(t_1))] \; [1 - \exp \{-(q_D \lambda/D\bar{\theta}_f) \; (t_2 - t_1)\}] \qquad (4)$$

Since q_D, the solution flux at depth D, is the only unknown in Eq. (4) it can be found. Then Eq. (4) may be used to predict $\bar{c}$ at any time t ($> t_1$), assuming the regime is the same as for the data that yielded the particular value of q_D. This allows judgement to be made on the necessity of changing irrigation practices to keep salt levels satisfactory in the long term.

MODEL FOR MOVEMENT OF A PEAK IN SOLUTE CONCENTRATION

The application of nitrogenous fertilizer, for example, typically results in the formation of a peak in nitrate concentration which, despite dispersive spreading, can be observed to move down the soil profile in response to leaching by water. The objective here is to develop and test a simple deterministic model of the leaching of nitrate (or other solute, assumed non-sorbed).

Mass continuity of water in the presence of a sink term s representing volumetric rate of uptake by plant roots requires that:

$$\frac{\partial \theta}{\partial t} = -\frac{\partial q}{\partial z} - s \qquad (5)$$

Combining the continuity equations for water (Eq. (5)), and solute (Eq. (1)), (assumed not taken up by the plant) gives:

$$\theta \frac{\partial c}{\partial t} = c \, s - q \frac{\partial c}{\partial z} \qquad (6)$$

Equation (6) will be used to describe movement of the peak of a solute concentration profile even though taken up by plants, thus assuming that spatial variability in uptake with depth leads to only minor distortion of peak solute position. Further assumptions are that the soil does not possess "large" cracks or fissures, and that following water entry, drainage to a characteristic 'field capacity' value takes place within a time period relatively short compared to that over which the solute peak position is to be followed. To simplify presentation, such a field capacity (denoted θ_{fc}), will be assumed constant with depth. This assumption (strictly that the water content range effectively available to plants is constant with depth) is adequate in the applications given.

The assumption of 'field capacity' behaviour enables a much-simplified but moderately accurate treatment of water dynamics in the broad class of soils displaying this behaviour, which include soils in which substantial leaching can take place. It allows emphasis to be placed on the end result (for solute as well as water), of dynamic transport processes, rather than on the details of such dynamics. This approach is also consistent with separating an infiltration process during which $\partial q/\partial z \gg s$ in Eq. (5) from periods of evapotranspiration, during which $s \gg \partial q/\partial z$.

During an infiltration process, putting $s = o$ in Eq. (6) gives:

$$\Theta \frac{\partial c}{\partial t} = - q \frac{\partial c}{\partial z} \tag{7}$$

Denote the value of z corresponding to the solute concentration peak by α . Since $c = f(z,t)$,

$$\frac{dc}{dz} = \frac{\partial c}{\partial z} + \frac{\partial c}{\partial t} \frac{dt}{dz} \tag{8}$$

At the position of a peak or maxima in concentration, neglecting the effect of dispersion (i.e., assuming peak concentration is constant):

$$dc/dz = o \tag{9}$$

Thus rate of movement of the peak position, $d\alpha/dt$, is given from Eqs. (8) and (9) with $z = \alpha$ by:

$$\frac{d\alpha}{dt} = - \frac{\partial c/\partial t}{\partial c/\partial z} \tag{10}$$

Substituting from Eq. (1) into Eq. (7), and assuming that during infiltration, effectively:

$$\theta = \Theta_{fc} \tag{11}$$

it follows that:

$$\frac{d\alpha}{dt} = \frac{q}{\Theta_{fc}} \tag{12}$$

This is a general equation describing movement of peak solute position (z), which is unaltered during an evapotranspiration period in which:

$$s \gg \partial q/\partial z \qquad (13)$$

During such periods, root water withdrawal (term s) leads to concentration increase, which follows from Eqs. (13) and (6) as:

$$\Theta \frac{\partial c}{\partial t} = c \, s \qquad (14)$$

If E is the rate of evapotranspiration, assumed uniform over the depth D_R of the root zone, then:

$$s = E/D_R \qquad (15)$$

and in general the rate of change of Θ (assumed uniform over the root zone) is given by:

$$\frac{\partial \Theta}{\partial t} = \frac{I - E}{D_R}, \quad (\Theta \leq \Theta_{fc}) \qquad (16)$$

where I is rate of infiltration. Thus the evaportranspirative process, whilst not directly responsible for displacement of the solute peak, is indirectly effective though modifying the water content prior to the subsequent infiltration event (Θ_{n-1} say), which in Figure 1 is illustrated as leading to a wetting front of depth z_n^*. Here subscripts n and n-1 represent finite calculational periods, the equivalent ponded depth of water infiltrating in period n being denoted by $I_n \Delta t$ (Figure 1(c)). This figure illustrates the basis of a practical computational prodecure based on the above theory presented in general (continuous) form (Dayananda et al. 1980; Rose et al. p.c.).

In Figure 1(c), ΔW_z is the depth of water which has passed depth z in time interval Δt. Hence $\Delta W_z = \int_0^{\Delta L} q \, dt$, or $q = \frac{dW_z}{dt}$, so that like ΔW_z, q in Eq. (12) is a linearly declining function of z reaching zero at $z = z_n^*$.

In Figure 1(a), the position z of the solute peak in the n'th computational interval is denoted α_n, having been displaced $\Delta \alpha_n$ from its previous position.

In Figure 1(b), the cross-hatched areas are equal to each other and to infiltration $I_n \Delta t$. It can be shown that Eq. (12) is consistent with considering the prior water described by Oagf in Figure 1(b) to be displaced completely ahead of the infiltrating water $I_n \Delta t$, ending up effectively in cdeg, these areas being equal.

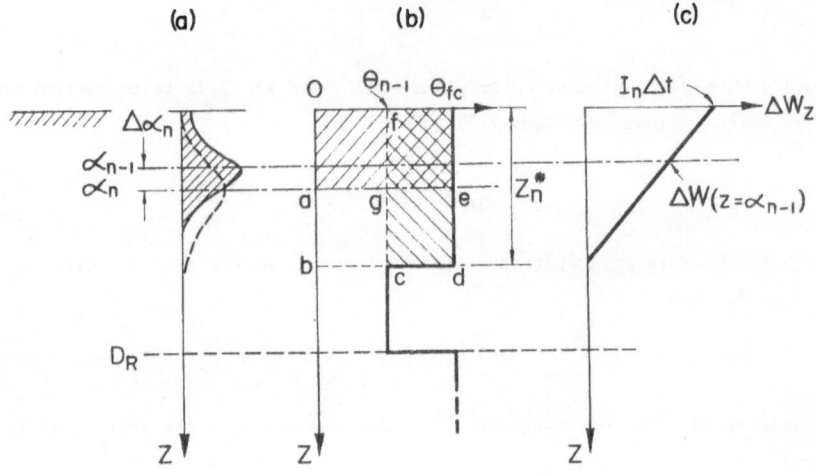

FIGURE 1. Basis of the model for movement of mean depth of penetration, α, of NO_3^- concentration peak in a "non-leaching period"; D_R, depth of root zone; z^*, depth of penetration of wetting front; θ, water content; I, infiltration rate, t, time; ΔW_z, equivalent ponded depth of water passing depth z .

Predictions based on the above type of theory were compared with observations by Saffigna et al. (1977) on the timing of emergence of concentration peaks in chloride and nitrate ions through 1.5 m of loamy sandy soil in Wisconsin. These concentration peaks resulted from surface applications of ammonium nitrate and potassium chloride to a potato crop, and were measured in a lysimeter. Both ions emerged at the same time. Table 1 shows agreement between the date of arrival at 1.5 m of observed and predicted peak concentrations.

In Table 1 the reporting by Saffigna et al. (1977) of the observed arrival of the concentration peak is mostly as a time period rather than a specific date. This time spread was due to uncertainty caused by dispersive spreading of the solutes, and to interference between concentrations resulting from successive applications of fertilizer and nitrate-rich irrigation water.

Dispersion as a process modifying solute concentrations in soil is quite

TABLE 1

Fertilizer Application Date (in 1973)	Reported date of observed concentration peak	Predicted date of arrival of concentration peak
25 April (planting)	31 May	Week 4 May
5 June	21 June – 24 July	Week 3 July
26 June	24 July – 13 August	Week 1 August
11 July	13 – 27 August	Week 2 August

Comparison of the timing of arrival at 1.5 m depth in soil of nitrate and chloride concentration peaks with that predicted by the type of model outlined in the text. (Data from Saffigna et al. 1977, treatment CON (1973)).

generally significant. A simple way of modelling such affects is now illustrated.

MODEL FOR THE EFFECTS OF DISPERSION ON SOLUTE CONCENTRATION

The differential equation describing solute transport, assuming (for the present) no uptake or transformation of the solute, but representing both the effects of mass flow (or convection) and dispersion is as follows:

$$\frac{\partial}{\partial t} (\Theta c) = \frac{\partial}{\partial z} (D\Theta\frac{\partial c}{\partial z}) - \frac{\partial}{\partial z} (qc) \qquad (17)$$

where D is the dispersion coefficient for solute in soil.

Wierenga (1977) and De Smedt and Wierenga (1978) have shown in laboratory columns that the concentration distribution resulting from irregular solute fluxes is very well approximated by the application of theory for an equivalent steady flux q, where:

$$q = const. \qquad (18)$$

Making similar assumptions for solute transport in soil under field conditions, and using Eq. (5) (with s = o), Eq. (17) can be simplified to:

$$\frac{\partial c}{\partial t} = D \frac{\partial^2 c}{\partial z^2} - V \frac{\partial c}{\partial z} \qquad (19)$$

where $V = q/\Theta$

= average pore water velocity

Following solution of applied fertilizer by rainfall and soil water, the resulting solute concentration may be approximated by a rectangular pulse of width ΔF, and initial concentration c_o, as shown in the top of Figure 2. When the solute is nitrate, the result of the generally rather rapid process of "immobilization" may be represented by subtracting the immobilised from the applied nitrogen before considering leaching of the remaining nitrate.

Let us now consider the effect of dispersion on the initially rectangular concentration pulse of Figure 2 as solute is leached vertically downward through soil.

For constant flux boundary conditions (Eq. 18), the change in relative concentration c/c_o, for a step input of concentration is given by De Smedt and Wierenga (1978) as:

$$\frac{c}{c_o} = 1/2 \ erfc \left[\frac{z - \alpha}{2 \ (D_o t + \varepsilon\alpha)^{0.5}} \right] \ . \qquad (20)$$

where α = mean solute penetration depth (cf. Figure 1),
z = depth from the initial position of the step change in concentration, here taken as depth from the soil surface,
D_o = molecular diffusion coefficient appropriate to movement in soil of the ion in question,
t = time from application of the step change in concentration,
and ε = dispersivity defined by:

$$D = D_o + \varepsilon V$$

In field soils the apparent value of D is affected by heterogeneity as well as by diffusion and hydrodynamic dispersion in the pore space. Thus, in practice, ε has to be experimentally determined from field data.

Dispersion of the rectangular concentration input of Figure 2 can be built up as indicated in the lower part of Figure 2 using the principle of superposition (Rose, 1977) to give:

$$\frac{c}{c_o} = 1/2 \left\{ \mathrm{erfc}\left(\frac{z - \alpha}{2(D_o t + \varepsilon\alpha)^{0.5}} \right) + \mathrm{erfc}\left(\frac{z - \beta}{2(D_o t + \varepsilon\beta)^{0.5}} \right) - 2 \right\} \tag{21}$$

where α is calculated as described in the previous section, and:

$$\beta = \alpha - \Delta F \tag{22}$$

It can be shown (Rose et al. p.c.), that provided solute uptake by plants is not too large compared to the solute being leached, the effect of such uptake can be approximated by multiplying the right hand side of Eq. (21) by a factor f such that mass of solute is conserved. Thus modified, Eq. (21) provides an "equivalent factored analytical profile" which approximately but analytically represents the net effects of convection, dispersion and sinks on the concentration profile.

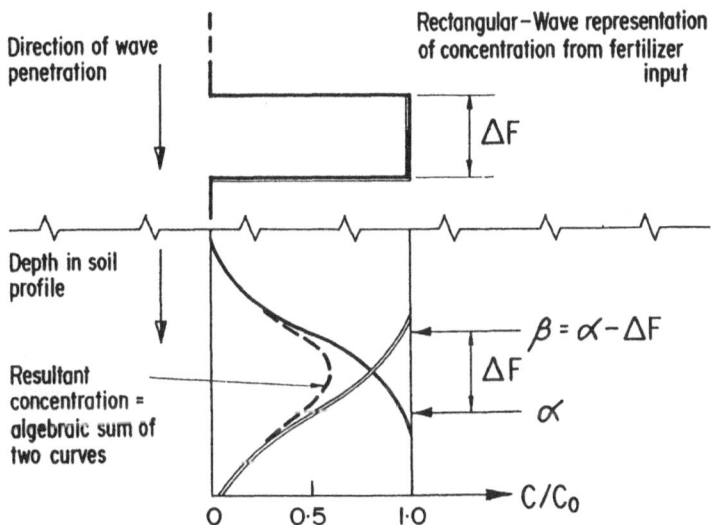

FIGURE 2. Dispersion accompanying convective transport of rectangular NO_3^- pulse of width ΔF; α, position of solute penetration depth; $\beta = \alpha - \Delta F$; c, solute concentration; c_o, initial solute concentration.

In summary, the approach to predicting solute concentration profiles as influenced by leaching and other (separately determined) uptake or transformation processes, is firstly to calculate the position α of the solute peak using the methods of the previous section. From Figure 2 the peak position is close to $(\alpha + \beta)/2$, but this is $\simeq \alpha$ if ΔF is small, as it commonly is.

Secondly, Eq. (21) (modified if needed by factor f discussed above) is then used to calculate the entire concentration profile, or the concentration at any particular depth of interest (z).

APPLICATION OF MODELS TO FIELD EXPERIMENTAL DATA

Application of this method is illustrated from the data of Chichester and Smith (1978), who measured the fate of ^{15}N-labelled fertilizer NO_3 applied during corn culture in four indisturbed (monolith) field lysimeters at Coshocton, Ohio, (of area 8 m^2) (Harrold and Dreibelbis, 1967). Topsoil was a silt loam, but from approximately 1.5 m to the measurement depth of 2.44 m, the material was decomposed siltstone or clay shale bedrock (Kelley et al. 1975). This bedrock was permanently fissured.

Curve c_D in Figure 3 shows the ^{15}N concentration measured in the lysimeter percolate. Curve c_B is the same concentration calculated at z = 2.44 m from the data of Chichester and Smith (1978) using the theory outlined in the two previous sections. This monthly calculation assumed uniform soil properties throughout the entire profile, despite the evidence of cracked weathered bedrock in the lower metre.

Discrepancies between measured (c_D) and calculated (c_B) concentrations in Figure 2 possibly may be explained in terms of the influence of the cracked bedrock material. This could lead to more rapid preferential crack movement in this lower layer. Rose et al. p.c. have presented a quantitative hypothesis that such preferential movement could be accompanied by diffusion of some 30 - 40 kg/ha of nitrogen into the porous bedrock when leachate concentration was increasing, to reappear (Figure 3) in later periods associated with the annual (northern hemisphere) winter flushing of the soil profile. Flushing of some ^{15}N mineralised in the previous summer may also be a minor contributor to the secondary maxima in measured concentration (Figure 3).

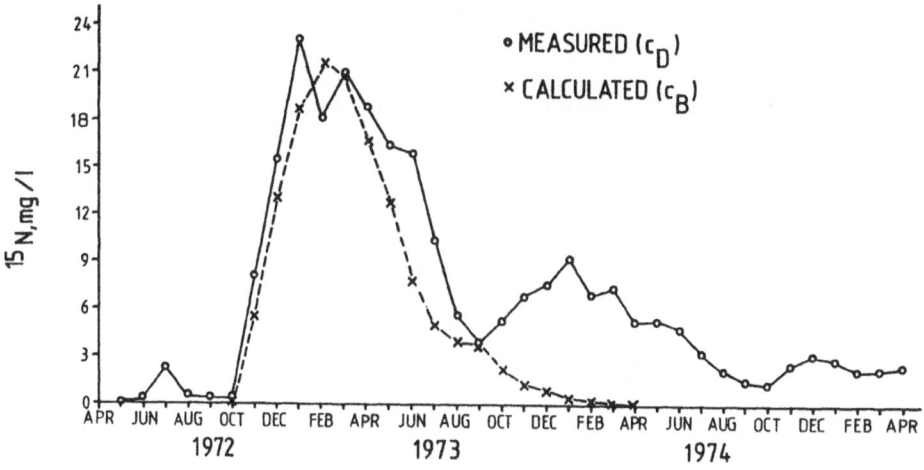

FIGURE 3. Calculated (c_B, x) vs. measured (C_D, o) mean monthly concentrations of ^{15}N (mg/l or ppm) in percolate from Coshocton lysimeters. Measured data from Chichester and Smith (1978) and calculations were at z = 244 cm (lysimeter depth), with ε = 12 cm and ΔF = 10 cm.

ACKNOWLEDGEMENTS

I gratefully acknowledge the stimulation of discussion with many colleagues, including Professor J. Y. Parlange, Dr. W. Chichester, Dr. J. Williams, Dr. J. T. Ritchie, Dr. P.W.A. Dayananda and Dr. W. Hogarth.

Mr. G. Sander provided valuable programming assistance.

REFERENCES

Chichester, F.W. and Smith, S.J. 1978. Disposition of ^{15}N-labelled fertilizer nitrate applied during corn culture in field lysimeters. J. Environ. Qual., 7, 227-233

Dayananda, P.W.A., Winteringham, F.P.W., Rose, C.W. and Parlange, J.Y. 1980. Leaching of a sorbed solute: a model for peak concentration displacement. Irrigation Sci., 1, 169-175

De Smedt, F. and Wierenga, P.J. 1978. Approximate analytical solution for solute flow during infiltration and redistribution. Soil Sci. Soc. Am. J., 42, 407-412

Harrold, L.L. and Driebelbis, F.R. 1967. Evaluation of agricultural hydrology by monolith lysimeters 1956-62. USDA Tech. Bull., 1367, 123 pp.

Kelley, G.E., Edwards, W.M., Harrold, L.L. and McGuinness, J.L. 1975. Soil of the North Appalachian Experimental Watershed. USDA Mics. Publ. No. 1296, 122 pp. illus.

Rose, D.A. 1977. Hydrodynamic dispersion in porous materials. Soil Science, 123, 277-283

Rose, C.W., Chichester, F.W., Williams, J.R., and Ritchie, J.T. (Personal communication). A contribution to simplified models of field solute transport.

Rose, C.W., Dayananda, P.W.A., Nielsen, D.R. and Biggar, J.W. 1979. Long-term solute dynamics and hydrology in irrigated slowly permeable soils. Irrigation Science, 1, 77-87

Saffigna, P.G., Keeney, D.R. and Tanner, C.B. 1977. Nitrogen, Chloride and Water Balance with Irrigated Russet Burbank Potatoes in a Sandy Soil. Agron. J., 69, 251-257

Wierenga, P.J. Solute distribution profiles computed with steady-state and transient water movement models. Soil Sci. Soc. Am. J., 41, 1050-1055.